THE PLANTS OF MOUNT KINABALU
I. FERNS AND FERN ALLIES

THE PLANTS OF MOUNT KINABALU

I. FERNS AND FERN ALLIES

BARBARA S. PARRIS

REED S. BEAMAN

JOHN H. BEAMAN

Royal Botanic Gardens
Kew

First published 1992

ISBN 0 947643 35 4

Addresses of authors:

Barbara S. Parris, 16A O'Neill's Ave., Takapuna, Auckland 9, New Zealand.
Reed S. Beaman, 820 NE 5th Ave., Gainesville, Florida 23601, U.S.A.
John H. Beaman, Department of Botany & Plant Pathology,
    Michigan State Univ., East Lansing, Michigan 48824, U.S.A.

General Editor of Series: J.M. Lock

Cover Design by Media Resources, RBG, Kew

Typeset at Royal Botanic Gardens, Kew by Pam Arnold,
Christine Beard, Brenda Carey, Margaret Newman,
Pam Rosen and Helen Ward

Printed & Bound in Great Britain by Whitstable Litho, Whitstable, Kent

CONTENTS

Professor R. E. Holttum on the occasion of his ninety-fifth birthday, 20 July 1990.

DEDICATION

In memory of Richard Eric Holttum (1895–1990), preeminent pteridologist. As Kinabalu towers over the whole of Borneo, Holttum's contributions to our knowledge of the ferns transcend this intellectual landscape.

INTRODUCTION

The present account comprises part of a botanical inventory or enumeration of the entire flora of vascular plants of Mount Kinabalu, the highest mountain in Borneo. This separate treatment of the pteridophytes was envisaged initially because we wanted to dedicate the work to Professor Holttum and also because it represents a logical subdivision of the large number of species comprising the total flora. The overall plan of the project was described by Beaman and Regalado (1989), and is followed here with minor modifications. These are discussed in the section on computer methods. Selection of the pteridophytes as the initial subject also has had the advantage of enabling us to follow a somewhat natural sequence for families, beginning with the fern allies and ferns, then the orchids also as a separate treatment, and finally the gymnosperms and remaining angiosperms.

"The Ferns of Mount Kinabalu" by Christensen and Holttum (1934) represents the first publication, other than the early studies of Stapf (1894) and Gibbs (1914), to attempt a comprehensive treatment of a major segment of the Kinabalu flora. The work of Christensen and Holttum sets a solid foundation for understanding of the fern flora and, even nearly 60 years later, illustrates a high standard of excellence in floristic work for a region still not well known.

The following enumeration lists 609 species of pteridophytes for Mount Kinabalu. These belong to 28 families and 145 genera, including almost all the principal genera of the Old World. An additional 12 infraspecific taxa bring the total number of pteridophyte taxa now known to 621. No new species are described in this account, but 10 new combinations for species and four for varieties are published here by Parris. The enumeration includes 30 taxa without published names, of which 29 will be described later by Parris and one to be described later by Chambers. Thirteen taxa are included for which there appear to be authentic records but for which we have seen no specimens.

We have attempted to cite type specimens only for taxa described from Mount Kinabalu. Likewise, the synonymy includes only names, and their types, that have been described from Mount Kinabalu. In addition to true synonyms, we also have noted synonymy sensu Christensen and Holttum. Likewise, names they recognized in the same sense as in this work are referenced in the text. Names of authors of taxa are abbreviated in accordance with the new list of author abbreviations being developed at Kew (Powell & Brummit (Eds.) in prep.).

In view of the extreme localization of many species (Beaman and Beaman 1990) and the fact that nearly one-third of the mountain has never been visited by a botanist, it appears probable that the list may be significantly extended in the future. The total number of pteridophytes for the world is about 12,000. It is remarkable, therefore, that approximately five percent of them occur on this one mountain in an area of only about 700 km^2. To make another comparison, there are more species of pteridophytes on Kinabalu than in the whole of mainland tropical Africa, which has about 500 species (Parris 1985).

While the Kinabalu flora is extremely rich, it must be noted that the records we have assembled represent over a century of collecting. It may therefore be unrealistic to believe that all the species recorded are there at the present time. The phenomenon of small-scale extinction (or extirpation) was noted by Parris after the drought of 1983. *Hymenophyllum peltatum* was present in 1988 only as large mats of dead fronds, whereas in 1980 it was green and lush. In Grammitidaceae the following species were seen in 1980, but were absent in 1985 and 1988 in spite of searching the same habitats and localities: *Acrosorus streptophyllus, A. friderici-et-pauli, Calymmodon muscoides, C. sp. 4, C. sp. 7, Ctenopteris denticulata, C. millefolia, C. subsecundodissecta, C. taxodioides, Grammitis caespitosa, G. friderici-et-pauli, G. jagoriana, G. oblanceolata, G. sp. 2, G. sp. 4, G. sp. 5,* and *G. sp. 6,* i.e., 17 of 76 species or about 22 percent local extinction within ten years. The local extinction may well be cyclic.

Collections of Pteridophytes from Mount Kinabalu

Christensen and Holttum (1934) enumerated 417 species from Kinabalu, about half of which were accounted for by new collections made by Holttum in 1931 and J. and M. S. Clemens in 1931-32. The Clemenses were still in the field when the Christensen and Holttum manuscript was completed. In that report Holttum summarized the work of previous collectors and cited earlier literature. No other floristic accounts of the pteridophytes of Kinabalu have been published subsequent to the Christensen and Holttum work. According to our analysis, their enumeration was based on 1364 collections, resulting from the work of nine collectors or collecting teams; 770 from J. and M. S. Clemens, 344 from Holttum, 122 by D. L. Topping, 56 by L. S. Gibbs, 29 by F. W. Burbidge, 23 by G. D. Haviland, 11 by H. Low, five by C. M. Enriquez, and four by G. A. G. Haslam.

The present report is based on nearly three times the number of specimens available to Christensen and Holttum, i.e., about 3500 collections including over 4400 specimens. These have been obtained by 44 collectors or collecting teams and are deposited in 14 herbaria. However, the great majority of specimens we have examined are in K and SING, particularly K, which has provided the primary resource for this inventory of pteridophytes.

Collectors for whom we have examined more than eight specimens are listed below with the approximate date of their time in the field and number of collections examined.

J. H. Beaman, October 1983–July 1984, August 1990: 171
F. W. Burbidge, 1877–78: 18
W. L. Chew, E. J. H. Corner & A. Stainton, June–September 1961: 206
W. L. Chew & E. J. H. Corner, January–April 1964: 19
M. S. Clemens, October–December 1915: 57
J. & M. S. Clemens, 1931–33: 1042
S. Collenette, 1960–66: 41

P. J. Edwards, March 1986: 51
H. P. Fuchs, August–September 1963: 18
L. S. Gibbs, February 1910: 27
G. D. Haviland, March–April 1892: 28
R. E. Holttum, November 1931: 353; November 1972: 22
M. Jacobs, October 1958: 36
S. Kokawa & M. Hotta, January–February 1969: 34
H. Low, March 1851, April–May, July 1858: 14
W. Meijer, 1959–1966: 37
B. Molesworth-Allen, February 1957: 19
B. S. Parris & J. P. Croxall, October–November 1980: 703
B. S. Parris, December 1985: 66; May 1988: 215
G. Shea & Aban Gibot, May 1973: 22
J. Sinclair, Kadim b. Tassim & Kapis b. Sisiron, June 1957: 56
D. L. Topping, October–November 1915: 142
Miscellaneous collections of 22 collectors: 69

Joseph and Mary Strong Clemens, who spent a total of over two years on Mount Kinabalu, including just over a month in 1915 and about two years in 1931–33, collected more pteridophytes on Mount Kinabalu than have any other collectors. They also obtained vast numbers of other kinds of plants, the entire collection probably totaling over 9000 numbers. Three collectors, D. L. Topping, R. E. Holttum, and B. S. Parris, concentrated on collecting ferns, and our knowledge of the fern flora has benefitted greatly from this specialist collecting. In about four months of field work Parris managed to collect almost as many pteridophytes as the Clemenses obtained in two years. The Kinabalu database includes 41 collections of Holttum that have been designated as types and 15 of Topping; 51 collections of the Clemenses are types.

Specimens obtained by the Royal Society expeditions are not cited below by collector but only by the prefix "RSNB." Collections of the first Royal Society expedition (1961) are by Chew, Corner & Stainton, and include numbers lower than 2980. Those of the second expedition (1964) are by Chew & Corner, and begin at 4000. A few numbers, beginning with 6000, were obtained by Chai & Ilias.

Elevation data given on labels in feet have been converted to meters; all elevations are rounded to the nearest 100 m. The stated elevations of 6000–13,500 ft for Clemens specimens from "Upper Kinabalu" have been ignored.

Diversity and Ecology of the Ferns[4]

An introduction to diversity and distribution patterns in the flora of Mount Kinabalu was published recently (Beaman and Beaman 1990). That paper attempts to explain the richness of the Kinabalu flora on the basis of an extensive range of climatic and edaphic conditions, rapid evolutionary rates, frequent speciation events, and long- and short-distance dispersal of plants pre-adapted to montane conditions. The level of endemism for pteridophytes appears to be much lower than for many of the groups of flowering plants. Rapid evolution and speciation may therefore be less important for them than for the latter. Among the 145 genera of pteridophytes represented in the Kinabalu flora, for example, only *Asplenium*, *Ctenopteris*, *Cyathea*, *Diplazium*, and *Grammitis* appear to be highly speciose, with occurrence in each genus in the area of 20 or more species. The spore habit of pteridophytes, of course, is particularly well adapted for long-distance dispersal. If the richness of the pteridophyte flora cannot be accounted for primarily by rapid rates of evolution and speciation, this may be well compensated by the highly effective dispersal ability of these plants.

In the wet tropics ferns are abundant everywhere but, as the great majority of them are not very large, they rarely dominate the landscape. They are adapted to various types of habitats afforded by trees and other large plants which form the bulk of the natural vegetation. Pteridophytes growing on the ground in forest shade are adapted to permanent weak light and high humidity. Where forest has been felled, or in open places such as river-banks, different species, which will tolerate strong light and (from time to time) drying winds, establish themselves; these sun ferns do not flourish in permanent shade.

Pteridophytes occur frequently as epiphytes in the wet tropics. They are adapted in various ways to their habitat, which involves exposure of their roots to the air and thus to frequent dry periods. On Mount Kinabalu about two-fifths of the total number of pteridophytes are epiphytes. Just over half the flora is made up of terrestrial species, and lithophytic species account for nearly ten percent. Tree-ferns, about four percent of the flora, form a special group, as their trunks can raise the crown of fronds some distance above the ground into less humid air and brighter light. Figure 1 provides an analysis of the elevational distribution by lifeform of the Mount Kinabalu pteridophyte taxa.

The greatest number of genera (98) and species (309) has been collected at about 1500 m, with gradually decreasing numbers known from above and below this level (Fig. 2). We believe that this graph reflects fairly accurately the actual diversity at the various elevations rather than being an artifact of collecting activity. Species from the lower elevations of Mount Kinabalu may be under-represented, however, because much of the lower elevation forests were destroyed before significant collecting activity had taken place there. Parris and Edwards (unpublished) have collated the data on pteridophytes collected by British botanists in the Danum valley (eastern Sabah) in lowland and hill dipterocarp forest between 150 and 550 m and record 152 taxa, 91, or 59 percent of which also occur on Mount Kinabalu. Similar diversity could be expected in the lowland forests of Mount Kinabalu. Data have been published by Jacobsen and Jacobsen (1989) for pteridophyte floras of southern and eastern Africa, also demonstrating the greatest diversity of species at around 1500 m.

---

[4] Modified from "The Ferns of Kinabalu National Park" (Holttum 1978).

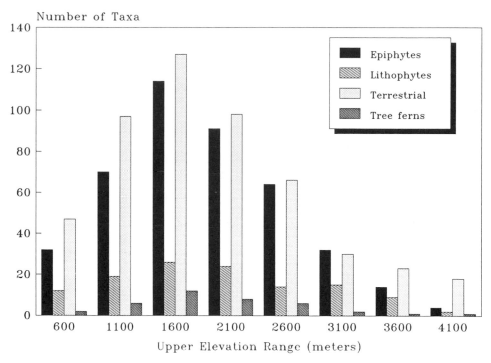

Fig. 1. Elevational distribution by life form of pteridophytes on Mount Kinabalu.

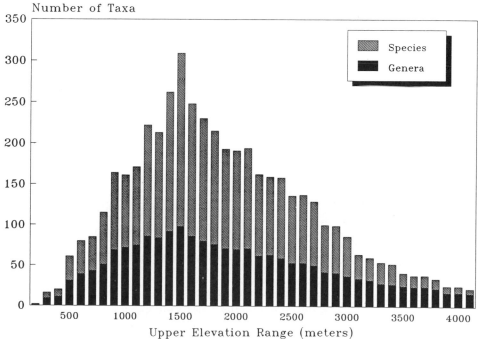

Fig. 2. Elevational distribution of pteridophyte taxa on Mount Kinabalu.

Roadside and Thicket-Ferns at Lower Elevations

The commonest low-elevation roadside fern is *Blechnum orientale* (Blechnaceae). Ironically, we know of only two collections of this species from Mount Kinabalu, only one of which we have seen. It has a tufted growth of pinnate fronds and is often conspicuous by the red colour of its young fronds. *Blechnum* is a genus of the southern hemisphere; in north temperate regions there is only one species. Its characteristic feature is that the sori lie along each side of the under surface of the midrib of each pinna. High on the mountain are other species of *Blechnum*, such as *B. fluviatile*, of very different appearance.

Among the other more conspicuous ferns along the approach roads to Kinabalu are members of the family Gleicheniaceae. These branch in a repeated forking pattern, sometimes symmetric and sometimes zigzag by alternate unequal forkings. In this family the main axis of the frond has a temporary dormancy, during which a pair of branches grows to its fullest development; only when this development is completed will the main axis resume growth. This feature enables the ferns to form dense and tangled thickets, as each frond is of indefinite growth. Higher up on the mountain, on exposed ridges and other open places, different members of the family occur, with different branch-patterns; some have only periodic dormancy of the main axis of fronds. There are also ferns of other families which have this type of periodic dormancy and can thus form thickets.

High-Elevation Thicket-Ferns

At about 1500 m the character of the thicket-forming sun-ferns changes, with increasing variety. In addition to the Gleicheniaceae, two different ferns form thickets by periodic dormancy of their main axes. One is *Hypolepis brooksiae* (Dennstaedtiaceae), with small ultimate segments, and its main branches covered with small prickles. The other is *Histiopteris stipulacea* (Dennstaedtiaceae), which has elongate simple pinnules with reflexed margins forming the indusia. On ridges at 2000 m and above *Diplopterygium bullatum* is frequent. It is very xeromorphic, with densely scaly young parts, the upper surface convex between veins. This bullate condition occurs frequently in many different groups of flowering plants, as well as ferns, particularly on ultramafic outcrops on Mount Kinabalu.

Tree-ferns

Large tree-ferns, often occurring in groups, can be seen near the roads in and around Kinabalu Park headquarters. Most of these belong to one species, *Cyathea contaminans* (Cyatheaceae), which is widely distributed in the Malesian region. The young stipes have a pale bluish waxy surface, with many light brown scales at their bases; the scales conceal sharp thorns. Tree-ferns of this kind occur up to about 1500 m, above which other species appear, mostly of smaller stature. In all, 22 species of tree-ferns of the genus *Cyathea* have been found on Kinabalu. All have a more or less persistent mass of scales protecting the young fronds; scale characters, both of these large basal scales and of smaller ones on upper parts of fronds, are important for distinguishing the species. Sori are sometimes protected by an indusium, sometimes not. One species (*C. tripinnata*) has more deeply divided fronds than the others, the ultimate segments being of the third order; its sori are covered by overlapping scales, not by a single indusium, although the latter condition was reported by earlier observers. It belongs to a small group of species in the eastern part of the Malesian region, extending to Queensland. The tree-fern *Dicksonia mollis* (Dicksoniaceae) occurs occasionally in forests at 1400 to 2400 m. Its young fronds are covered with a dense mass of reddish hairs, some thick and bristle-like, not flat scales as in *Cyathea*. It is known only from Borneo and the Philippines. Related species occur in New Guinea and Australasia.

Ferns of Cultivated Areas

The giant fern *Cibotium arachnoideum* (Dicksoniaceae), with a massive prostrate rhizome instead of a trunk, has fronds even larger (4-5 m long) than tree-ferns. Its rhizome is covered with soft yellow-brown hairs which are used for staunching bleeding wounds. It survives burning when land is cleared for cultivation and persists on steep ladangs (fields) at 900-1200 m. Also in areas cleared for cultivation, but not conspicuous elsewhere, is *Pteridium esculentum* (Dennstaedtiaceae), related to the bracken fern of north temperate areas. It has a thick creeping rhizome some distance below the ground which is not killed by slash-and-burn agriculture. Members of the Gleicheniaceae, in contrast, have a more slender rhizome, creeping on or near the ground surface, and are readily destroyed by burning.

Shade Ferns of the Forest

Inside the forest, especially at about 1500-2000 m, one finds many terrestrial ferns among the smaller plants of the forest floor. The most abundant are members of the large family Thelypteridaceae, which have tufted fronds with more or less deeply lobed pinnae and small round sori on the lower surface. These are similar in general habit, but have many differences in detail; 52 species are known from Kinabalu. Of similar habit, but sometimes with bipinnate fronds, are plants of the genus *Diplazium* (Woodsiaceae), which have elongated sori along the veins, protected when young by thin, narrow indusia. Ferns with broader, less deeply lobed pinnae, an elaborate network of veins, and small round sori belong to the genus *Tectaria* (Dryopteridaceae). Occasionally near streams one may find very large plants of *Angiopteris* (Marattiaceae), which may be recognized by the swollen fleshy bases of their stipes, and by sori consisting of rather large sporangia of unusual form along the veins near the edges of pinnae. This genus and its allies are very different in sporangia and vegetative anatomy from most other ferns. They are considered to be related to some Palaeozoic fossils, but the relationship is not very certain. Some small ground ferns of several genera, including *Asplenium* and *Lindsaea*, are further discussed below.

In the wet forests of middle elevations ferns are seen not infrequently climbing tree trunks. These have rhizomes up to 1 cm thick that are attached to the bark of trees. The genus *Teratophyllum* (Lomariopsidaceae) is particularly noteworthy among the climbers. The sterile fronds of these ferns have simple elongate pinnae, jointed to the main axis of the frond. The fertile fronds, not often found, have very narrow pinnae, the lower side of which are densely covered with sporangia. Young plants of *Teratophyllum* have small fronds (bathyphylls) with many pinnae of a shape different from those on old plants; they may sometimes be seen on the lower parts of tree-trunks.

On rocks by streams in shady situations other ferns of several different genera are found. Some have differentiated sterile and fertile fronds, the fertile

7

with narrower pinnae. The venation patterns are critical for distinguishing these taxa. At lower elevations some species of *Asplenium* (Aspleniaceae) grow in such places; their elongate sori along the veins, with rather firm translucent indusia, are distinctive, although fronds may vary considerably in different species, from simple to variously branched.

On the main summit trail on the ridge below Kemburongoh two species of *Lindsaea* (Dennstaedtiaceae) are distinctive ground ferns in the forest. The fronds are about 0.3 m high, with few branches, each bearing several small pinnae. The sori are near the edges of the pinnae and are protected by indusia attached below the sori, open toward the edge of the pinnae. *Lindsaea* species are all rather small; the genus occurs throughout the wetter parts of the tropics but with different species in different regions. *Lindsaea* and its allies are a rather isolated group of ferns and undoubtedly old, but little is known of their fossil history.

Ferns of Mountain Ridges

In more open places along the Kemburongoh ridge are two genera, *Dipteris* (Dipteridaceae) and *Matonia* (Matoniaceae) which do have fossil relatives that were once almost world-wide, dating from the Jurassic. Both have fronds of a peculiar branching pattern which results in an almost circular outline, the main veins bearing successively smaller branches on their lower sides only. The divisions of the fronds of *Dipteris* are broad, with an elaborate network of veins; the main veins fork in their distal parts. Sporangia are borne in small sori on the lower surface, without any indusia. *Dipteris* occurs only in the Malesian region and western Pacific, with isolated species in northeastern India and southern China. In addition to the mountain species (two on Kinabalu), two others grow beside rocky streams in forests in the low country locally in Borneo, Sumatra, and Peninsular Malaysia.

Matonia has successively smaller branches on the lower side, each branch arising near the base of the previous one. The individual branches of the frond are 30 cm or more long and deeply lobed, rigid in texture, and glossy dark green on the upper surface. The sori on the lower surface are protected by thick peltate indusia. This fern is found locally on exposed high ridges in Borneo, Peninsular Malaysia, Sumatra and New Guinea. The plants spread by means of a creeping rhizome and thus form colonies; young plants grown from spores are rarely seen.

In more or less shaded places on the ridges from 1800 m upward are three other peculiar fern genera. A few species of the genus *Coryphopteris* (Thelypteridaceae) have small trunks like miniature tree-ferns but have simply pinnate fronds only 0.2 to 0.4 m long. Allied species are found in leached peat or sandy soil near the crests of mountain ridges throughout Malesia, about 40 species in all, with the greatest abundance in New Guinea; a few also occur in northeastern India and southern China. Ferns of the genus *Plagiogyria* (Plagiogyriaceae) have tufted, simply pinnate fronds with entire pinnae. The sterile and fertile fronds are dimorphic, the latter having very narrow pinnae. Their rhizomes and stipes usually lack hairs or scales, but the croziers are covered with slime which may take over the protective function of scales. White pneumatophores project through the slime. The third fern, *Cheiropleuria bicuspis* (Cheiropleuriaceae), is very different, having simple rigid brittle pale green fronds on slender stipes, the fertile much narrower than the sterile. Sterile fronds sometimes have a simple pointed apex, sometimes two points more or less widely

Plate 1. **A** Eastern shoulder of Mount
Kinabalu from the Pinosuk
Plateau near the West Mesilau
River at about 1500 m.

B The glaciated rock face of the
southern summit area from
Panar Laban at about 3400 m.

C *Cyathea havilandii* (Cyatheaceae)
in open areas of stunted
*Dacrydium gibbsiae/Leptospermum
recurvum* forest on the southern
slope near Layang-layang at
about 2600 m.

Plate 2. **A** *Matonia pectinata* (Matoniaceae).

B *Belvisia squamata* var. *borneensis* (Polypodiaceae).

C *Coryphopteris badia* (Thelypteridaceae).

D *Parathelypteris beddomei* var. *beddomei* (Thelypteridaceae).

E *Pleuromanes pallidum* (Hymenophyllaceae).

F *Lecanopteris pumila* (Polypodiaceae).

Plate 3. **A** *Diplopterygium brevipinnulum*
(Gleicheniaceae).

B *Dicranopteris curranii*
(Gleicheniaceae).

C *Acrophorus nodosus*
(Dryopteridaceae).

D *Diplopterygium bullatum*
(Gleicheniaceae).

E *Sticherus truncatus* var. *truncatus*
(Gleicheniaceae)

F *Microtrichomanes ridleyi*
(Hymenophyllaceae).

Plate 4. **A** *Lycopodiella cernua*
(Lycopodiaceae).

B *Huperzia serrata*
(Lycopodiaceae).

C *Cheiropleuria bicuspis*
(Cheiropleuriaceae).

D *Lindsaea repens* var. *sessilis*
(Dennstaedtiaceae).

E *Dipteris novoguineensis*
(Dipteridaceae)

F *Dipteris conjugata*
(Dipteridaceae).

separated, but fertile fronds are always simple. The creeping rhizome has soft hairs.

In more open places on the Kemburongoh ridge are plants of *Blechnum vestitum*, which has narrow fertile pinnae much like those of *Plagiogyria*. The young fronds, both sterile and fertile, are bright red. Unlike *Plagiogyria*, however, they have many scales when young. The species is closely related to taxa in Australia and South Africa.

Ferns of High Elevations

At about 2900 m a small tree-fern, *Cyathea havilandii*, with a short trunk (sometimes no evident trunk) and all fronds densely scaly, is common on the ridges. In open places on the higher ridges a very peculiar little fern, *Schizaea fistulosa* (Schizaeaceae), looking like a dwarf rush, with a tuft of small sporangia-bearing branches at the top of each slender stipe, is fairly common. It has a distribution from Peninsular Malaysia to New Caledonia and Fiji. Other species of *Schizaea* occur at lower elevations.

In the valley forest at 3000 m, where the soil is much richer than on the ridges, some fine terrestrial ferns of temperate-latitude types occur, including a *Polystichum* (Dryopteridaceae) with much-branched fronds 1 m or more high, bearing small rounded pinnae and sori protected by circular indusia. A true *Dryopteris* (Dryopteridaceae), *D. wallichiana*, with reniform indusia, also occurs in this habitat.

In the summit area some ferns may be found growing under projecting rocks which shelter them from the sun. These include species of *Asplenium* (Aspleniaceae), with elongate sori covered by indusia, and *Athyrium* (Woodsiaceae), with sori of varying shape from reniform to J-shaped or linear. They are all rather small plants and have more scleromorphic fronds than usually are found in these genera. Also in such situations are very small ferns of the Grammitidaceae, with simple fronds. This group is discussed below, in the section on high-elevation epiphytes, as most of them are epiphytic.

A small species of *Blechnum*, *B. fluviatile*, with pinnae only about 2 cm long, occurs among rocks in more or less sheltered places at 2700 m or more. It is known otherwise only from New Guinea and Australasia.

Epiphytic Ferns

Epiphytic ferns are abundant on almost all large trees at all elevations. At lower elevations those on the branches of tall trees may be seen with binoculars or on fallen limbs or trees. Low-elevation epiphytes also occur near the ground on smaller trees on open river banks. Epiphytic ferns are very varied, but belong mainly to the families Aspleniaceae, Davalliaceae, and Polypodiaceae. At high elevations, species of the Grammitidaceae also are abundant.

Members of the Polypodiaceae have a creeping, usually scaly rhizome to which the stipe of each frond is jointed. When the frond is old it falls and leaves a smooth scar. The life-span of a single frond of an epiphytic fern is yet to be observed in Malaysia. The fronds are almost always tough or fleshy. They are simple, or at most once pinnate, and thick enough to store some water; a thick cuticle retards water loss. Although most of the Polypodiaceae are relatively small, *Aglaomorpha heraclea* has fronds up to 2 m long.

All Polypodiaceae have sporangia in round or elongate sori with no indusia. Their veins form an elaborate network, with branching small free veins in the interstices. The fronds, however, are often so thick that the veins cannot be readily seen. Genera of the Polypodiaceae are classified partly on the nature of the scales of the rhizome and partly on sori and venation. One genus, *Belvisia*, has simple fronds with the sporangia contiguous on a narrow apical portion, the sporangia being protected when young by small imbricate scales. Another peculiar member of the family is *Lecanopteris*. Its fleshy rhizome develops internal hollows that are often colonized by ants. The ants carry small seeds to their nest as food, some of which germinate between the crevices in the old parts of the rhizome which dies away at the base while the tip keeps on growing. A colony of epiphytes is thus formed which gradually increases in bulk, the old dead parts of each plant decaying and helping form a mass of humus that acts as a sponge holding rain water. Detritus from the ants also provides nutrients for the colony of epiphytes.

Ferns of the family Davalliaceae, in contrast to the Polypodiaceae, have much-divided fronds with small ultimate segments. Their creeping rhizome with jointed frond bases, however, is similar to that of the Polypodiaceae. At elevations of 900 to 1500 m species of *Davallia* and *Davallodes* may be found as epiphytes, or sometimes on rocks in more open places. The fronds bear pinnae which are progressively smaller from base to apex, and the individual pinnae have asymmetric bases. In *Davallia* the scales are uniformly thin, and the sori are protected by indusia attached at the base and sides (like a pocket sewn on a coat), open toward the margin of the frond. The shape of the sori and the rhizome scales vary from species to species. Plants of *Davallodes*, known only from along streams, have scales with a small flat base which bears a dark bristle 1 cm or more long. The form of indusia in *Davallodes* varies; one Kinabalu species (*D. borneense*) has almost reniform indusia rather like *Thelypteris* or *Dryopteris*, the other (*D. burbidgei*) has indusia exactly as in *Davallia*. *Davallodes* also differs from *Davallia* in the branching pattern of the fronds, which are always elongate and narrowed toward the base, whereas *Davallia* fronds are usually widest at the base. A third genus with one species, *Araiostegia hymenophylloides*, occurs as a low-level epiphyte in shaded places from about 1400 to 2100 m. Its rhizome and fronds are similar to *Davallia*, but the frond is thinner and the indusium is attached only at the base of the sorus. At higher elevations plants of the genus *Humata*, similar to *Davallia* but smaller and often more xeromorphic, are epiphytic in rather open places. As in *Araiostegia*, the indusia are attached at the base of the sorus.

Plants of some species of *Asplenium* grow on rocks by streams or on the ground in forests at mid elevations, but more are epiphytes; 37 species of the genus are known from Kinabalu. The commonest species at lower elevations is the bird-nest fern, *A. nidus*, with simple broad fronds to 1 m or more long and over 10 cm wide. The sori can be seen on the basal half of each of the closely parallel lateral veins. These plants catch fallen leaves from trees and hold them. The mass of decaying leaves held between the bases of the fronds and the roots of the fern form a sponge which holds water and can form the base of a colony of other epiphytes. At the other extreme, some species have bipinnate or even tripinnate fronds which may hang downwards, the largest having fronds 1-1.5 m long.

An interesting genus of epiphytes which always has simple fronds with free veins is *Elaphoglossum* (Lomariopsidaceae). The fertile fronds, usually narrower than the sterile ones and sometimes on longer stipes, are completely covered beneath with sporangia that lack indusia. The fertile fronds are probably produced seasonally, perhaps in response to dry weather. *Elaphoglossum* species have a short, creeping, scaly rhizome which has outgrowths 1-2 cm long to which the fronds are imperfectly jointed. Species differ in the details of the scales on the rhizome and fronds (which always have distinctive minute scales on their surfaces), in the shape of the fronds, and the relative shape of fertile and sterile fronds. In the whole Malesian region 50 species of *Elaphoglossum* are known; eight species are recorded from Mount Kinabalu. Where *Elaphoglossum* plants grow with other epiphytes, their fronds at a distance look much like the leaves of orchids, so they are easy to overlook.

The Vittariaceae are represented by 17 species on Kinabalu. They are epiphytes or lithophytes with simple fronds. *Vittaria* fronds are relatively long and narrow, and the larger ones are pendulous. Sporangia are produced either in marginal to submarginal grooves or in a narrow band on each side of the midrib; these are accompanied by short paraphyses which have peculiar enlarged reddish or brown terminal cells. *Antrophyum* species have broader fronds, with sporangia in grooves along the veins, forming a more or less continuous network. They also have sporangia with characteristic hairs, differing in the various species. The smallest plants of the family belong to the genus *Vaginularia*. These have thread-like fronds 5–10 cm long.

Filmy Ferns

Filmy ferns (Hymenophyllaceae) occur abundantly at higher elevations, where clouds gather nightly so that the mass of epiphytes on trees is dripping wet for a considerable part of each 24 hours. At lower elevations they are confined to the most humid places in the forest, often near streams. These ferns have the peculiar character that each frond, apart from the veins, consists of a single layer of cells. Because of their leaf structure Hymenophyllaceae can grow only where the air is very humid most of the time, although many of them can withstand some drying for short periods if not exposed to strong sunlight and drying winds. On Kinabalu 57 species of the family have been collected.

Most Hymenophyllaceae have a slender creeping rhizome with small fronds spaced along it, but the larger ones have fronds in a tufted arrangement, the larger 30 cm or more long. Broadly speaking, there are two types of sori in the filmy ferns. In the large genus *Hymenophyllum*, represented on Kinabalu by the segregate genera *Hymenophyllum*, *Mecodium*, *Meringium*, *Microtrichomanes*, and *Sphaerocionium*, each sorus is protected by two equal outgrowths at the base of the receptacle. The large genus *Trichomanes*, represented here by the segregate genera *Callistopteris*, *Cephalomanes*, *Crepidomanes*, *Gonocormus*, *Macroglena*, *Microgonium*, *Nesopteris*, *Pleuromanes*, *Reediella*, *Selenodesmium*, and *Vandenboschia*, in contrast, has the receptacle growing in the base of a funnel or short tube, the mouth of which may be dilated or sometimes two-lipped. These broad divisions are not sharp, and much diversity occurs in *Hymenophyllum* sens. lat. and *Trichomanes*. For these reasons, Copeland's segregate genera are distinguished here.

High-Elevation Epiphytes

The most abundant group of epiphytic ferns on the higher parts of the mountain are in the family Grammitidaceae. They also grow on rocks in rather open places. Altogether 76 species are known from Kinabalu, many occurring from 2000 m upward. Over one-third belong to the genus *Grammitis*, with simple fronds 3-30 cm long and 0.5-2 cm wide, with free (usually branched) veins. They nearly always have slender stiff hairs (usually reddish) on their stipes, and often on the fronds themselves. The sori are usually round and have no indusia. Thus, they have some of the characters of the Polypodiaceae, and in the past most have been included in the genus *Polypodium*. Nevertheless, they are a distinct group which merits family rank.

The relationship of the Grammitidaceae to other fern families is uncertain. Apart from *Grammitis* itself, most other members of the family have more or less deeply lobed simple fronds, and differ in the nature of their hairs, scales, venation, and position of the sori. In some cases the sporangia have stiff little hairs which can be seen by careful examination with a hand lens. One genus of the family, *Scleroglossum*, has the sporangia developed in shallow grooves near the edges of the small, simple, narrow fronds. They resemble *Vittaria* superficially, but are very different in frond structure, spores, and scales.

Comparison of the Pteridophyte Flora of Mount Kinabalu with Other Areas

In general, the pteridophytes at altitudes up to 1000 m on Mount Kinabalu and its foothills are also found in Peninsular Malaysia and Sumatra, and many also in the Philippines. Above 1000 m a considerable proportion of species have more restricted distributions, many being confined to Borneo, although closely related species often occur in Peninsular Malaysia. At elevations over 2000 m are some species with a wider distribution to the east and south, to the mountains of New Guinea and to New Zealand. The latter distribution pattern is also known in the gymnosperms and some genera of flowering plants of high levels on Kinabalu. A few high-elevation pteridophytes are related to species of north temperate regions. A relatively small number (ca. 50; i.e. 8 percent) of the Kinabalu pteridophytes appear to be restricted to Mount Kinabalu. Three-quarters of these are terrestrial, one-eighth are tree-ferns, and the rest lithophytic and epiphytic.

The nearest flora to that of Mount Kinabalu is that on the remainder of the Crocker Range and a comparison of their pteridophytes is interesting. Parris (unpublished, on the basis of her field work and examination of herbarium specimens) records 157 pteridophytes occurring between 300 m and 1800 m in the Crocker Range in lowland and hill forest. Of these, 140 or 89 percent are also found on Mount Kinabalu. Between the same elevations on Kinabalu 520 pteridophytes are known, pointing up its extreme richness in comparison with the Crocker Range.

Perhaps the most thoroughly studied pteridophyte flora in Borneo is that of Gunung (Mount) Mulu in Sarawak. The Pteridophyta of Gunung Mulu were enumerated by Parris ct al. (1984), where the authors suggest that about 425 species occur, but additional information (Parris unpublished) suggests that 416 taxa is a more realistic figure, i.e., about 180 fewer than on Mount Kinabalu. The areal extent of Mulu is somewhat smaller and the maximum elevation much lower (i.e., 2377 m vs. 4101 m), which accounts for its smaller flora, although with extensive alluvial and limestone forests it has a high diversity of habitats. Over a similar elevational range (300 to 2377 m) Mulu has 399 taxa, 290 or 73 percent of which are shared with Mount Kinabalu, and Kinabalu has 549 taxa, i.e., 53 percent in common with Mulu. Such similarities are to be expected on other peaks of similar height in Borneo, although data are lacking.

Three peaks of fairly similar height in Papua New Guinea also merit comparison. They are Mount Wilhelm (4509 m), Mount Hagen (4000 m) and Mount Giluwe (4088 m) [elevations from the *Times Atlas*]. Of the 621 taxa recorded on Mount Kinabalu, 73 are known from the montane areas of one or more of these three peaks and 20 taxa occur on all three. Johns and Stevens (1971) recognise 110 species of pteridophytes occurring above 2743 m on Mount Wilhelm, to which Parris (unpublished) has added another 24, from subsequent collections and re-identification of those cited by Johns and Stevens. Above 2700 m on Mount Kinabalu 142 taxa of pteridophytes are known. For Mount Hagen and Mount Giluwe figures are available for ferns only (Parris unpublished; based on field work and examination of herbarium specimens). The totals of *fern* taxa above 2700 m on the four peaks are as follows: Mount Kinabalu, 134; Mount Giluwe, 128; Mount Hagen, 124; Mount Wilhelm, 122. At lower elevations comparative figures are not available for Mount Wilhelm, but above 2400 m, 178 ferns are known from Mount Kinabalu, 140 for Mount Giluwe, and 151 for Mount Hagen. Data are lacking from New Guinea for further comparisons of the pteridophyte floras at lower elevations.

Mount Maquiling (1090 m) in Luzon, Philippines, has ca. 300 pteridophyte taxa (M. G. Price pers. comm.). This mountain is the most thoroughly examined area in the Philippines for pteridophytes, yet only half as many species are known from there as from Mount Kinabalu. The relatively low elevation of Mount Maquiling undoubtedly explains its significantly smaller pteridophyte flora. It should be noted that at or below 1100 m on Mount Kinabalu we have recorded only 259 taxa.

Computer Methods

The basic procedures outlined by Beaman and Regalado (1989) for development and management of a microcomputer specimen-oriented database have been followed in producing the present inventory of pteridophytes. Advances in computer technology since that paper was completed have facilitated our work. For example, we are now using *dBASE IV* as a database management environment, which has significantly speeded application programming and data processing. Although the Kinabalu database has been developed in an MS-DOS environment, it is now available on an *Ingres* Unix network platform using the Internet in the United States and comparable networks in Europe, Asia, Australia, and New Zealand.

Effective retrieval of information about the occurrence of taxa requires that localities be expressed in a standard manner. Variant spellings and names for many localities have presented a challenge in the standardisation process and in the preparation of a location map for Mount Kinabalu. We therefore have worked with personnel who are native speakers of the Dusun language but who also speak excellent English in an effort to standardise spellings of the localities. The standardised localities will be published in a gazetteer of place names (Beaman et al. in prep.), and will be shown on a location map now in preparation. We also anticipate producing a synonymised list of all locality expressions that have been recorded for plant specimen data.

Taxonomic Arrangement

The enumeration follows the list of families that the herbaria of the Royal Botanic Gardens, Kew, Royal Botanic Garden, Edinburgh, and the Natural

13

History Museum, London, have agreed to recognise. We use an alphabetical sequence, first of the fern allies, then the ferns, for families, genera, species, and infraspecific taxa. The definition of families of ferns has been in flux with only moderate consensus as to the family into which certain genera are placed. The placement of genera in families and in some cases the decisions as to what genera should be recognised follow essentially the system used at Kew.

ENUMERATION OF TAXA

Kramer, K. U. & Green, P. S., eds. 1990. The Families and Genera of Vascular Plants. Kubitzki, K., ed. I. Pteridophytes and Gymnosperms. Springer-Verlag, Berlin etc.; Pteridophytes, pp. 11–277. Tryon, A. F. & Lugardon, B. 1990. Spores of the Pteridophyta. Springer-Verlag, Berlin etc.

FERN ALLIES

1. EQUISETACEAE

1.1. EQUISETUM

Hauke, R. L. 1963. A taxonomic monograph of the genus *Equisetum* subgenus *Hippochaete*. Beih. Nova Hedwigia 8: 1–123.

1.1.1. Equisetum ramosissimum Desf.

a. subsp. **debile** (Roxb.) Hauke, Amer. Fern J. 52: 33 (1962).

Terrestrial. Open, wet areas. Elevation: 800–1500 m.

Material examined: DALLAS: 900 m, *Clemens 27420* (K), 900 m, *Holttum SFN 25276* (K); KADAMAIAN RIVER: 800 m, *Haviland 1353* (K); KIAU: *Low s.n.* (K); KUNDASANG/RANAU: 1100 m, *Sinclair et al. 8960* (K); MINITINDUK GORGE: *Clemens 10443* (K); PARK HEADQUARTERS: 1500 m, *Beaman 10908* (MSC).

2. LYCOPODIACEAE

Øllgaard, B. 1989. Index of the Lycopodiaceae. Biol. Skrifter 34: 1–135. Parris, B. S., Jermy, A. C., Camus, J. M. & Paul, A. M. 1984. The Pteridophyta of Gunung Mulu National Park. *In* Studies on the Flora of Gunung Mulu National Park, Sarawak. Ed. A. C. Jermy: 172–173. Forest Dept., Kuching, Sarawak. Tagawa, M. & Iwatsuki, K. 1979. Flora of Thailand. Vol. 3, Part 1: 7–13.

2.1. HUPERZIA

2.1.1. Huperzia australiana (Herter) Holub, Folia Geobot. Phytotax. 20: 70 (1985).

Lycopodium australianum Herter, Bot. Jahrb. Syst. 43: Beibl. 98: 42 (1909).

Terrestrial and lithophytic. *Dacrydium/Leptospermum* forest on ultramafic and granite, ridges and stream banks. Elevation: 2000–4100 m.

Material examined: DACHANG: 2700 m, *Clemens 28746* (K); EASTERN SHOULDER: 3100 m, *RSNB 727* (K), 2900 m, *921* (K); GURULAU SPUR: 2900 m, *Clemens 50803* (K); KADAMAIAN RIVER: 2000 m, *Meijer SAN 29117* (K); KING GEORGE PEAK: 2700–4000 m, *RSNB 5981* (K); LOW'S PEAK: 4100 m, *Clemens 27088* (K), 4000 m, *27775* (K); MOUNT KINABALU: 3400 m, *Haviland 1410* (K), 3400 m, *1411* (K), 3400–4000 m, *Holttum SFN 25489* (K); PAKA-PAKA CAVE: 3500 m, *Molesworth-Allen 3291* (K); PANAR LABAN: 3300 m, *Parris 11669* (K); SUMMIT TRAIL: *Clemens 10613* (K), 2600 m, *Parris 11668* (K); VICTORIA PEAK: 3800 m, *Clemens 51394* (K).

2.1.2. Huperzia phlegmaria (L.) Rothm., Feddes Repert. 54: 62 (1944).

15

Lycopodium phlegmaria L., Sp. Pl. 1101 (1753).

Epiphyte. Primary forest on ultramafic substrate. Elevation: 800–2300 m.

Material examined: DALLAS: 900 m, *Clemens 26404* (K); GOLF COURSE SITE: 1700–1800 m, *Beaman 7189* (MSC); HEMPUEN HILL: 800 m, *Abbe et al. 9930* (K); MAMUT COPPER MINE: 1600–1700 m, *Beaman 9942* (MSC); PIG HILL: 2000–2300 m, *Beaman 9872* (MSC).

2.1.3. Huperzia phyllantha (Hook. & Arn.) Holub, Folia Geobot. Phytotax. 20: 75 (1985).

Lycopodium phyllanthum Hook. & Arn., Bot. Beechey Voy., 102 (1832).

Epiphyte. Lower montane forest. Elevation: 1500–1700 m.

Material examined: PENIBUKAN: 1500 m, *Clemens 50319* (K); PINOSUK PLATEAU: 1700 m, *RSNB 1887* (K).

2.1.4. Huperzia pinifolia Trevis., Atti Soc. Ital. Sci. Nat. 17: 247 (1874).

Lycopodium piscium (Herter) Tagawa & K. Iwats. Acta Phytotax. Geobot. 22: 103 (1967).

Epiphytic and lithophytic. Primary lower montane and mossy scrub forest, sometimes on ultramafic substrates. Elevation: 1200–2300 m.

Material examined: GOLF COURSE SITE: 1700–1800 m, *Beaman 10662* (MSC); MAMUT RIVER: 1200 m, *RSNB 1741* (K); PIG HILL: 2000–2300 m, *Beaman 9873* (MSC).

2.1.5. Huperzia serrata (Thunb. ex Murray) Trevis., Atti Soc. Ital. Sci. Nat. 17: 248 (1874).

Lycopodium serratum Thunb. ex Murray, Syst. Veg. ed. 14, 944 (1784).

Terrestrial. Usually in lower montane oak forest. Elevation: 700–3000 m. The elevation data for *Clemens 29065* may be in error.

Material examined: DALLAS: 900 m, *Clemens 27758* (K); EASTERN SHOULDER: 800 m, *RSNB 593* (K); MARAI PARAI SPUR: 1800–2400 m, *Clemens 32957* (K); MELANGKAP KAPA: 700–1000 m, *Beaman 8803* (MSC); MESILAU BASIN: 3000 m, *Clemens 29065* (K); MESILAU CAVE TRAIL: 1700–1900 m, *Beaman 7978* (MSC), 1700–1900 m, *9117* (MSC); MESILAU RIVER: 1500 m, *RSNB 4301* (K), 1800–2400 m, *Clemens 51357* (K); PARK HEADQUARTERS: 1500 m, *Parris 11429* (K); SUMMIT TRAIL: 2200 m, *Parris 11561* (K); TENOMPOK: 1500 m, *Clemens s.n.* (K); TENOMPOK RIDGE: 1400–1500 m, *Beaman 8210* (MSC); TINEKUK RIVER: 1200 m, *Haviland 1416* (K); WEST MESILAU RIVER: 1600–1700 m, *Beaman 8666* (MSC), 1600 m, *9027* (MSC).

2.1.6. Huperzia squarrosa (G. Forster) Trevis., Atti Soc. Ital. Sci. Nat. 17: 247 (1874).

Lycopodium squarrosum G. Forster, Fl. Ins. Aust., 479 (1786).

Epiphytic and lithophytic. Lower montane forest, sometimes on ultramafic substrates. Elevation: 800–1700 m.

Material examined: LOHAN RIVER: 800–1000 m, *Beaman 9050* (MSC); LUBANG: *Holttum SFN 25563* (K); PENIBUKAN: 1200 m, *Clemens 32184* (K); PINOSUK PLATEAU: 1600 m, *Beaman 10796* (MSC), 1700 m, *RSNB 1889* (K); TAHUBANG FALLS: 1500 m, *Clemens 40314* (K).

2.1.7. Huperzia verticillata (L. f.) Trevis., Atti Soc. Ital. Sci. Nat. 17: 248 (1874).

Lycopodium verticillatum L. f., Suppl., pl. 448 (1782).

Epiphytic and lithophytic. Lower montane oak forest, sometimes on ultramafic substrates. Elevation: 1500–2400 m.

Material examined: BAMBANGAN CAMP: 1500 m, *RSNB 4582* (K); EAST MESILAU RIVER: 2000 m, *Collenette 21661* (K); GOLF COURSE SITE: 1800 m, *Beaman 7476* (MSC); MARAI PARAI SPUR: *Holttum SFN 25611* (K); MENTEKI RIVER: 1600 m, *Beaman 10789* (MSC); MESILAU BASIN: 2100–2400 m, *Clemens 29698* (K); MESILAU CAVE: 2000–2100 m, *Beaman 8135* (KNP), 1700–2000 m, *10678* (KNP); MESILAU CAVE TRAIL: 1700–1900 m, *Beaman 9116* (MSC); PINOSUK PLATEAU: 1700 m, *RSNB 1814* (K); WEST MESILAU RIVER: 1600 m, *Beaman 9026* (MSC).

2.2. LYCOPODIELLA

2.2.1. Lycopodiella cernua (L.) Pichi Serm., Webbia 23: 166 (1968).

Lycopodium cernuum L., Sp. Pl., 1103 (1753).

Terrestrial. Lower montane oak/chestnut forest. Elevation: 1500–1700 m.

Material examined: MEMPENING TRAIL: 1600 m, *Parris 11510* (K); MURU-TURA RIDGE: 1700 m, *Clemens 34030* (K); PARK HEADQUARTERS: *Parris 85/47* (K); TENOMPOK: 1500 m, *Clemens 28738* (K).

2.3. LYCOPODIUM

2.3.1. Lycopodium casuarinoides Spring, Bull. Acad. Roy. Sci. Belg. 8: 521 (1841).

Terrestrial. Scandent in low ridge forest. Elevation: 1500–2700 m.

Material examined: EASTERN SHOULDER: 2400 m, *RSNB 187* (K), 2700 m, *911* (K); GURULAU SPUR: 2100–2700 m, *Clemens 50789* (K); KEMBURONGOH: *Clemens 10520* (K), 2100 m, *Price 238* (K); KUNDASANG: 1500–1800 m, *Meijer SAN 23842* (K); MOUNT KINABALU: 1800 m, *Burbidge s.n.* (K), 2700 m, *Haviland 1414* (K), *Low s.n.* (K); SUMMIT TRAIL: 2000 m, *Beaman 6772* (MSC); TAHUBANG RIVER HEAD: 2100 m, *Clemens 32925* (K).

2.3.2. Lycopodium clavatum L., Sp. Pl., 1101 (1753).

Lycopodium kinabaluense Ching, Acta Bot. Yunnanica 4: 223 (1982). Type: Bambangan River: 1500 m, *RSNB 4413* (holotype PE n.v.; isotype K!).

Terrestrial. Lower montane and upper montane forest, particularly margins. Elevation: 1500–3500 m.

Additional material examined: EASTERN SHOULDER: 3100 m, *RSNB 725* (K); KEMBURONGOH/LUMU-LUMU: *Holttum SFN 25455* (K); MAMUT COPPER MINE; 1600–1700 m, *Beaman 9964* (MSC); MESILAU CAVE: 2400 m, *Clemens 51629* (K); MOUNT KINABALU: 1800 m, *Saikeh SAN 82754* (K); PAKA-PAKA CAVE: 3500 m, *Molesworth-Allen 3264* (K); PAKA-PAKA CAVE/PANAR LABAN: 3200 m, *Sinclair et al. 9139* (K); PANAR LABAN: 3400 m, *Weber SAN 54729* (K); PARK HEADQUARTERS/POWER STATION: *Aban SAN 79577* (K); SHANGRI LA VALLEY: 3400 m, *Collenette 21507* (K); SUMMIT TRAIL: 2600 m, *Parris 11603* (K); TENOMPOK: 1500 m, *Clemens 28532* (K), 1500 m, *29003* (K).

2.3.3. Lycopodium platyrhizoma Wilce, Nova Hedwigia 3: 99, t. 2, f. 5–6, t. 6 (1961).

Terrestrial. Low ridge forest. Elevation: 2400–2900 m.

Material examined: JANET'S HALT: 2400 m, *Collenette 561* (K); MENTEKI RIDGE: 2400 m, *RSNB 7139* (K); MESILAU RIVER: 2900 m, *Clemens 51510* (K).

2.3.4. Lycopodium scariosum G. Forster, Fl. Ins. Austr., 87 (1786).

Terrestrial. Upper montane ridge forest. Elevation: 2400–3400 m.

Material examined: EASTERN SHOULDER: 3100 m, *RSNB 726* (K); JANET'S HALT: 2400 m, *Collenette 544* (K); MESILAU BASIN: 3000 m, *Clemens 29405* (K); MOUNT KINABALU: 3400 m,

Haviland 1413 (K), 2600 m, *Meijer SAN 22048* (K), 2700 m, *29252* (K); PAKA-PAKA CAVE: *Holttum SFN 25511* (K), 2900 m, *Sinclair et al. 9196* (K); SHANGRI LA VALLEY: 3400 m, *Collenette 21510* (K); SHEILA'S PLATEAU/SHANGRI LA VALLEY: 3400 m, *Fuchs & Collenette 21444* (K).

2.3.5. Lycopodium volubile G. Forster, Fl. Ins. Austr., 86 (1786).

Terrestrial. Lower and upper montane forest, may be scrambling in low vegetation. Elevation: 2100–2700 m.

Material examined: EASTERN SHOULDER: 2700 m, *RSNB 908* (K); JANET'S HALT/SHEILA'S PLATEAU: *Collenette 21561* (K); KEMBURONGOH: 2100 m, *Sinclair et al. 9046* (K); MESILAU TRAIL: 2400 m, *Meijer SAN 38586* (K); TAHUBANG RIVER HEAD: 2100 m, *Clemens 32925* (K).

2.3.6. Lycopodium wightianum Wall. ex Grev. & Hook., Bot. Misc. 2: 379 (1831).

Terrestrial. Lower and upper montane forest, one record on open wet slope. Elevation: 2400–3400 m.

Material examined: GURULAU SPUR: 2400 m, *Clemens 50928* (K); KEMBURONGOH/PAKA-PAKA CAVE: 2400–3400 m, *Clemens 28980* (K); MOUNT KINABALU: 3400 m, *Haviland 1412* (K); SUMMIT TRAIL: 2600 m, *Parris 11604* (K); UPPER KINABALU: *Clemens 28861* (K).

3. PSILOTACEAE

3.1. PSILOTUM

Tagawa, M. & Iwatsuki, K. 1979. Flora of Thailand. Vol. 3, Part 1: 5–6.

3.1.1. Psilotum nudum (L.) P. Beauv., Prodr. Aetheogam., 112 (1805).

Terrestrial. Ridge forest, one record on ultramafic substrate. Elevation: 1100–1500 m.

Material examined: HEMPUEN HILL: 1100 m, *Abbe et al. 9955* (K); TENOMPOK: 1500 m, *Clemens 27988* (K).

4. SELAGINELLACEAE

4.1. SELAGINELLA

Parris, B. S., Jermy, A. C., Camus, J. M. & Paul, A. M. 1984. The Pteridophyta of Gunung Mulu National Park. *In* Studies on the Flora of Gunung Mulu National Park, Sarawak. Ed. A. C. Jermy: 173–175. Forest Dept., Kuching, Sarawak. Tagawa, M. & Iwatsuki, K. 1979. Flora of Thailand. Vol. 3, Part 1: 14–31. Wong, K. M. 1983. Critical observations on Peninsular Malaysian *Selaginella*. Gardens' Bull. 35: 107–135.

4.1.1. Selaginella biformis A. Braun, Forsch. Gaz. 4, Bot. 6: 17 (1889).

Elevation: 1400 m.

Material examined: MOUNT KINABALU: 1400 m, *Clemens 27514* (K).

4.1.2. Selaginella bluuensis Alderw., Bull. Jard. Bot. Buit. 2, 11: 29 (1913).

Terrestrial. Secondary hill forest. Elevation: 1000 m.

Material examined: KIAU/TAHUBANG RIVER: 1000 m, *Parris 11530* (K).

4.1.3. Selaginella brevipes A. Braun, App. Ind. Sem. Hort. Berol., 1 (1867).

Terrestrial. Lower montane forest, hill forest, lowland forest. Elevation: 400–1500 m.

Material examined: DALLAS: 900 m, *Clemens 28154* (K), 900 m, *Holttum SFN 25252* (K); KAUNG/DALLAS: 400–1000 m, *Holttum SFN 25133* (K); PARK HEADQUARTERS: 1400 m, *Edwards 2163* (K), *2185* (K), 1500 m, *Parris 11424* (K); TENOMPOK: 1400 m, *Holttum SFN 25397* (K); TENOMPOK/LUMU-LUMU: *Holttum SFN 25436* (K).

4.1.4. Selaginella conferta T. Moore, Proc. Roy. Hort. Soc. 1: 133 (1861).

Terrestrial. Secondary hill forest. Elevation: 400–1000 m.

Material examined: DALLAS: 800 m, *Holttum SFN 25367* (K); KAUNG/DALLAS: 400–1000 m, *Holttum SFN 25131* (K).

4.1.5. Selaginella delicatula (Desv.) Alston, J. Bot. 70: 282 (1932).

Terrestrial. Hill and lower montane forest, open areas. Elevation: 600–1800 m.

Material examined: DALLAS: 600 m, *Holttum SFN 25368* (K); LIWAGU RIVER TRAIL: 1400 m, *Edwards 2180* (K); MINITINDUK GORGE: 900 m, *Clemens 29623* (K); ULAR HILL TRAIL: 1800 m, *Parris 11460* (K).

4.1.6. Selaginella dielsii Hieron., Hedwigia 51: 254 (1911) [1912].

Terrestrial. Probably on ultramafic substrate. Elevation: 1500 m.

Material examined: MARAI PARAI: 1500 m, *Holttum SFN 25620* (K).

4.1.7. Selaginella furcillifolia Hieron., Hedwigia 50: 31 (1910).

Terrestrial. Hill forest. Elevation: 900 m.

Material examined: DALLAS: 900 m, *Clemens 26650* (K).

4.1.8. Selaginella hewittii Hieron., Hedwigia 51: 262 (1911) [1912].

Terrestrial. Hill forest. Elevation: 900–1000 m.

Material examined: DALLAS: 900 m, *Clemens 27208* (K), 1000 m, *Holttum SFN 25262* (K).

4.1.9. Selaginella ingens Alston, J. Bot. 72: 229 (1934). Type: DALLAS: 900 m, *Clemens 27020* (holotype BM n.v.; isotype K!).

Terrestrial. Lowland forest and hill forest. Elevation: 900 m. Not endemic to Mount Kinabalu.

4.1.10. Selaginella intermedia (Blume) Spring, Bull. Acad. Brux. 10: 144 (1843).

a. var. **intermedia**

Terrestrial. Hill forest. Elevation: 900 m.

Material examined: DALLAS: 900 m, *Clemens 27600* (K).

4.1.11. Selaginella involvens (Sw.) Spring, Bull. Acad. Brux. 10: 136 (1843).

Terrestrial and epiphytic. Lowland forest and hill forest. Elevation: 800–2100 m.

Material examined: DALLAS: 800 m, *Holttum SFN 25361* (K); KAUNG: *Burbidge s.n.* (K); MESILAU BASIN: 2100 m, *Clemens 29048* (K); TAHUBANG RIVER: 900 m, *Haviland 1415* (K).

4.1.12. Selaginella ornata (Hook. & Grev.) Spring, Bull. Acad. Brux. 10: 232 (1843).

Epiphytic and lithophytic. Lower montane valley forest. Elevation: 1600 m.

Material examined: LIWAGU RIVER TRAIL: 1600 m, *Parris 11485* (K); TENOMPOK: *Holttum SFN 25398* (K).

4.1.13. Selaginella aff. paxii Hieron.

Terrestrial. Lower montane valley oak forest. Elevation: 1500 m. Endemic to Mount Kinabalu.

Material examined: PARK HEADQUARTERS: 1500 m, *Parris 85/46* (K), 1500 m, *11430* (K).

4.1.14. Selaginella rugulosa Ces., Atti della R. Acad. Sci. 7(8): 35 (1876).

Terrestrial, fronds prostrate. Hill forest and lower montane forest, locally common, sometimes on ultramafics. Elevation: 1000–1500 m.

Material examined: KIAU/TAHUBANG RIVER: 1000 m, *Parris 11529* (K); PARK HEADQUARTERS: 1500 m, *Parris 85/45* (K), 1500 m, *11428* (K); PENIBUKAN: 1300 m, *Parris 11521* (K).

FERNS

Holttum, R. E. 1959. General keys to Pteropsida. Fl. Males. 2, 1 (1): IX–XXI.

5. ADIANTACEAE

5.1. ADIANTUM

Copeland, E. B. 1958. Fern Flora of the Philippines. Vol. 1: 158–164. Holttum, R. E. 1968. Revised Flora of Malaya. Vol. 2, Ferns, Ed. 2: 596–604.

5.1.1. Adiantum diaphanum Blume, Enum. Pl. Jav., 215 (1828). C. Chr. & Holttum, Gardens' Bull. 7: 285 (1934).

Terrestrial. Lower montane forest, often near water. Elevation: 900–2100 m.

Material examined: DALLAS/TENOMPOK: 1200 m, *Holttum SFN 25290* (K, SING); KIAU/LUBANG: *Topping 1608* (SING); KILEMBUN RIVER HEAD: 1400 m, *Clemens 32521* (K); LUBANG: *Clemens 10327* (K), *10358* (K), 1200 m, *Holttum SFN 25553* (K, SING); LUMU-LUMU: 2100 m, *Clemens 29462* (K); MAMUT COPPER MINE: 1400 m, *Collenette 1034* (K); MINITINDUK GORGE: 900–1500 m, *Clemens 26982* (K); PINOSUK PLATEAU: 1500 m, *Parris & Croxall 9138* (K); TENOMPOK: 1500 m, *Clemens 27513* (K), 1500 m, *28360* (K), 1500 m, *29506* (K).

5.1.2. Adiantum hosei Baker, J. Bot. 26: 324 (1888).

Terrestrial. Hill dipterocarp forest. Elevation: 1000 m.

Material examined: PORING HOT SPRINGS/LANGANAN WATER FALLS: 1000 m, *Parris & Croxall 8926* (K); UPPER KINABALU: *Clemens 30841* (SING).

5.1.3. Adiantum opacum Copel., Philipp. J. Sci. 1, Suppl.: 255, pl. 3 (1906).

Terrestrial. Low stature hill forest on ultramafic substrate. Elevation: 800–1000 m.

Material examined: LOHAN RIVER: 800–1000 m, *Beaman 9054* (K, MICH, MSC).

5.2. CHEILANTHES

Holttum, R. E. 1968. Revised Flora of Malaya. Vol. 2, Ferns. Ed. 2: 589–592.

5.2.1. Cheilanthes tenuifolia (N. L. Burm.) Sw., Syn. Fil., 129, 332 (1806). C. Chr. & Holttum, Gardens' Bull. 7: 285 (1934).

Terrestrial. Lowland. Elevation: 300 m.

Material examined: KEBAYAU/KAUNG: 300 m, *Clemens 27678* (K).

5.3. CONIOGRAMME

Copeland, E. B. 1958. Fern Flora of the Philippines. Vol. 1: 148–149. Holttum, R. E. 1968. Revised Flora of Malaya. Vol. 2, Ferns. Ed. 2: 588–589.

5.3.1. Coniogramme fraxinea (D. Don) Diels, Nat. Pfl. 1(4): 262 (1899). C. Chr. & Holttum, Gardens' Bull. 7: 284 (1934).

Terrestrial. Hill forest and lower montane forest by streams. Elevation: 1100–1500 m.

Material examined: DALLAS: 1100 m, *Holttum SFN 25277* (SING); KINATEKI RIVER: 1500 m, *Clemens 32930* (L, MICH); KINATEKI RIVER HEAD: 1500 m, *Clemens 32365* (L, MICH).

5.3.2. Coniogramme macrophylla (Blume) Hieron., Hedwigia 57: 291 (1916).

Coniogramme macrophylla (Blume) Hieron. var. *copelandii* (Christ) Hieron., Hedwigia 57: 291 (1916). C. Chr. & Holttum, Gardens' Bull. 7: 284 (1934).

Terrestrial with long fronds. Hill forest in ravines. Elevation: 800–900 m.

Material examined: DALLAS: 800 m, *Clemens 27227* (US); MINITINDUK: 900 m, *Holttum SFN 25588* (SING).

5.4. MONACHOSORUM

Copeland, E. B. 1958. Fern Flora of the Philippines. Vol. 1: 99–100. Holttum, R. E. 1968. Revised Flora of Malaya. Vol. 2, Ferns. Ed. 2: 633.

5.4.1. Monachosorum subdigitatum (Blume) Kuhn, Chaetopt., 345 (1882). C. Chr. & Holttum, Gardens' Bull. 7: 226 (1934).

Terrestrial. Usually in lower montane oak forest. Elevation: 1700–3000 m.

Material examined: EASTERN SHOULDER: 2000 m, *RSNB 147* (K, SING), 2000 m, *180* (K, SING), 2900 m, *795* (K, SING); GOKING'S VALLEY: 2700 m, *Fuchs 21470* (K); JANET'S HALT/ SHEILA'S PLATEAU: 2900 m, *Collenette 21548* (K); KEMBURONGOH: 1800 m, *Holttum SFN 25544* (SING); LIWAGU RIVER TRAIL: 1700 m, *Parris 11475* (K); MARAI PARAI: 3000 m, *Clemens 33209* (K), 3000 m, *33211* (K); MARAI PARAI SPUR: 1800–2100 m, *Gibbs 4092* (K); MESILAU CAVE TRAIL: 1700–1900 m, *Beaman 9100* (MICH, MSC); MOUNT KINABALU: 1800 m, *Low s.n.* (K); SUMMIT TRAIL: 2200 m, *Parris & Croxall 8662* (K); UPPER KINABALU: *Clemens 29938* (K), *29967* (K).

5.5. PITYROGRAMMA

Holttum, R. E. 1968. Revised Flora of Malaya. Vol. 2, Ferns. Ed. 2: 592–594.

5.5.1. Pityrogramma calomelanos (L.) Link, Handb. Gewachs. 3: 20 (1833). C. Chr. & Holttum, Gardens' Bull. 7: 284 (1934).

Terrestrial. Open situations. Elevation: 500–900 m.

Material examined: DALLAS: 900 m, *Clemens 27036* (K); TAKUTAN: 500 m, *Shea & Aban SAN 77117* (K).

5.6. SYNGRAMMA

Holttum, R. E. 1968. Revised Flora of Malaya. Vol. 2, Ferns. Ed. 2: 579–585.

5.6.1. Syngramma alismifolia (C. Presl) J. Sm., London J. Bot. 4: 168, t. 7–8, f. B (1845). C. Chr. & Holttum, Gardens' Bull. 7: 284 (1934).

Terrestrial. Hill dipterocarp forest. Elevation: 1000–1200 m.

Material examined: MARAI PARAI SPUR: 1200 m, *Clemens 32484* (K), *Topping 1868* (SING); PORING HOT SPRINGS/LANGANAN WATER FALLS: 1000 m, *Parris & Croxall 8928* (K).

5.7. TAENITIS

Holttum, R. E. 1968. A re-definition of the fern-genus *Taenitis* Willd. Blumea 16: 87–95.

5.7.1. Taenitis blechnoides (Willd.) Sw., Syn. Fil. 24, 220 (1806). C. Chr. & Holttum, Gardens' Bull. 7: 239 (1934).

Terrestrial. Hill dipterocarp and lower montane forest, often on ultramafic substrate. Elevation: 900–1400 m.

Material examined: DALLAS: 900 m, *Clemens 27218* (K); EASTERN SHOULDER: 1100 m, *RSNB 590* (K, SING); MELANGKAP TOMIS: 900–1000 m, *Beaman 8610* (K, MICH, MSC), 900–1000 m, *8986* (MICH, MSC); PENIBUKAN: 1200 m, *Clemens 31305* (K); PINOSUK PLATEAU: 1400 m, *Beaman 10730* (K, MICH, MSC), 1400 m, *10733a* (MICH); PORING HOT SPRINGS/LANGANAN WATER FALLS: 1000 m, *Parris & Croxall 8960* (K).

5.7.2. Taenitis hookeri (C. Chr.) Holttum, Kew Bull. 30: 334 (1975).

Terrestrial. Lower montane forest. Elevation: 1500 m.

Material examined: EASTERN SHOULDER: 1500 m, *RSNB 1568* (K, SING).

6. ASPLENIACEAE

6.1. ASPLENIUM

Copeland, E. B. 1960. Fern Flora of the Philippines. Vol. 3: 429–453. Holttum, R. E. 1968. Revised Flora of Malaya. Vol. 2, Ferns. Ed. 2: 413–443. Holttum, R. E. 1974. *Asplenium* Linn., sect. *Thamnopteris* Presl. Gardens' Bull. 27: 143–154. Parris, B. S., Jermy, A. C., Camus, J. M. & Paul, A. M. 1984. The Pteridophyta of Gunung Mulu National Park. *In* Studies on the Flora of Gunung Mulu National Park, Sarawak. Ed. A. C. Jermy: 214–215, Forest Dept., Kuching, Sarawak.

6.1.1. Asplenium affine Sw., Schrad. J. Bot. 1800 (2): 56 (1801).

Asplenium spathulinum J. Sm. ex Hook. var *amaurolobum* Rosenstock ex C. Chr. in C. Chr. & Holttum, Gardens' Bull. 7: 282 (1934). Type: Minitinduk: 900 m, *Holttum SFN 25582* (holotype BM n.v.; isotypes K!, SING!). *Asplenium spathulinum* J. Sm. ex Hook., Sp. Fil. 3: 170 (1860). C. Chr. & Holttum, Gardens' Bull. 7: 282 (1934).

Epiphyte. Hill forest and lower montane forest. Elevation: 800–1500 m.

Additional material examined: DALLAS: 900 m, *Clemens 26924* (US), 900 m, *28038* (K), 800 m, *Holttum SFN 25364* (K, SING); TENOMPOK: 1500 m, *Clemens 27769* (K, MICH, US), 1500 m, *28689* (K, MICH, SING); TENOMPOK/LUMU-LUMU: *Clemens 29418* (MICH).

6.1.2. Asplenium allenae Parris, ined.

Lithophytic. Low upper montane forest of *Rhododendron*, *Dacrydium* and *Leptospermum*. Elevation: 3300–3500 m. Not endemic to Mount Kinabalu.

Material examined: PAKA-PAKA CAVE: 3500 m, *Molesworth-Allen 3273* (K); PANAR LABAN: 3300 m, *Parris 11587* (K); SUMMIT TRAIL: 3300 m, *Parris & Croxall 8743* (K).

6.1.3. Asplenium amboinense Willd., Sp. Pl. 5: 303 (1810).

Lower montane forest.

Material examined: KIAU/LUBANG: *Topping 1815* (SING, US); LUBANG: *Topping 1798* (SING, US).

6.1.4. Asplenium batuense Alderw.

a. f. angusta ? C. Chr. in C. Chr. & Holttum, Gardens' Bull. 7: 275 (1934). Type: Kiau/Lubang: *Topping 1593* (holotype BM n.v.).

Hill forest? Based on *Topping 1593*, Kiau/Lubang (BM? n.v.). Endemic to Mount Kinabalu?

6.1.5. Asplenium borneense Hook., Sp. Fil. 3: 135, t. 186 (1860). C. Chr. & Holttum, Gardens' Bull. 7: 278 (1934). Type: Tahubang River: *Low s.n.* (holotype K!).

Terrestrial, epiphytic and lithophytic. Hill forest and lower montane forest by streams. Elevation: 800–2200 m. Not endemic to Mount Kinabalu.

Additional material examined: DALLAS: 800 m, *Holttum SFN 25365* (K, SING); KIAU/TAHUBANG RIVER: 1000 m, *Parris 11531* (K); MARAI PARAI SPUR: *Clemens 10916* (K), *Topping 1862* (SING, US); MESILAU CAVE: 1900–2200 m, *Beaman 9538* (MICH, MSC); PENATARAN BASIN: 1500 m, *Clemens 32565* (US); PENIBUKAN: 1200 m, *Clemens 31309* (K); SAYAP: 800–1000 m, *Beaman 9809* (K, MICH, MSC); TENOMPOK: 1500 m, *Clemens 28364* (K); TINEKUK RIVER: 1000 m, *Haviland 1477* (K).

6.1.6. Asplenium caudatum G. Forster, Prodr., 80 (1786).

Epiphytic and lithophytic. Lower montane oak forest, often on ultramafic substrate. Elevation: 1600–2200 m.

Material examined: MESILAU CAVE: 1900–2200 m, *Beaman 9536* (K, MICH, MSC), 1700–2000 m, *10679* (K, MICH, MSC); WEST MESILAU RIVER: 1600–1700 m, *Beaman 8669* (K, MICH, MSC), 1600–1700 m, *8685* (K, MICH, MSC), 1600 m, *9003* (MICH, MSC).

6.1.7. Asplenium cheilosorum Kunze ex Mett., Aspl., 133, n. 104, t. 5, f. 12–13 (1859).

Lithophytic and terrestrial. Lower montane oak forest by streams in deep shade. Elevation: 1500–1600 m.

Material examined: LIWAGU RIVER TRAIL: 1500 m, *Parris 11487* (K), 1600 m, *Parris & Croxall 9098* (K).

6.1.8. Asplenium cymbifolium Christ, Bull. Boiss. II, 6: 999 (1906).

Asplenium nidus sensu C.Chr. & Holttum non L., Gardens' Bull. 7: 725 (1934) p.p.

Epiphyte. Lower montane oak forest. Elevation: 1500 m.

Material examined: PARK HEADQUARTERS: 1500 m, *Parris 10838* (K), 1500 m, *Parris & Croxall 9027* (K); TENOMPOK: 1500 m, *Clemens 28577* (K, US).

6.1.9. Asplenium dichotomum Hook., Sp. Fil. 3: 210 (1860). C. Chr. & Holttum, Gardens' Bull. 7: 283 (1934). Type: Mount Kinabalu: *Low s.n.* (syntype K!).

Epiphyte. Lower montane forest. Elevation: 1100–1500 m. Not endemic to Mount Kinabalu.

Additional material examined: DALLAS/TENOMPOK: 1100 m, *Clemens 27737* (K); KADAMAIAN RIVER: 1500 m, *Haviland 1494* (K); KIAU/LUBANG: *Topping 1598* (US); LUBANG: 1200 m, *Holttum SFN 25551* (K); MAMUT RIVER: 1200 m, *RSNB 1651* (K); MESILAU CAMP: 1500 m, *RSNB 6005* (K); MOUNT KINABALU: *Burbidge s.n.* (K); PINOSUK PLATEAU: 1500 m, *Parris & Croxall 9136* (K); TENOMPOK: 1500 m, *Clemens 29305* (K).

6.1.10. Asplenium elmeri Christ, Philipp. J. Sci. 2C: 164 (1907). C. Chr. & Holttum, Gardens' Bull. 7: 283 (1934).

Lithophytic, epiphytic and terrestrial. Lower and upper montane forest. Elevation: 1500–4100 m.

Material examined: EASTERN SHOULDER: 2900 m, *RSNB 791* (K); GOKING'S VALLEY: 2700 m, *Fuchs 21468* (K); GOLF COURSE SITE: 1700–2000 m, *Beaman 9608* (K, MICH, MSC); KADAMAIAN RIVER HEAD: 3200 m, *Clemens 50897* (US); KEMBURONGOH/PAKA-PAKA CAVE: 3000 m, *Clemens 27057* (K); KILEMBUN RIVER: 2700 m, *Clemens 33744* (K); LAYANG-LAYANG: 2500 m, *Edwards 2202* (K); LOW'S PEAK: 4100 m, *Clemens 27046* (K); LUBANG/PAKA-PAKA CAVE: *Topping 1737* (US); MENTEKI RIDGE: 2900 m, *Collenette 21546* (K); MESILAU BASIN: 2100–2400 m, *Clemens 29711* (SING); MOUNT KINABALU: *Haviland 1480* (K); PAKA-PAKA CAVE: 3000 m, *Clemens 27964* (K, SING), 3000 m, *Holttum SFN 25502* (K, SING), *Molesworth-Allen 3303* (US), 3100 m, *Parris 11571* (K), 3000 m, *Parris & Croxall 8728* (K), 3000 m, *Sinclair et al. 9117* (K, SING), *Topping 1677* (SING, US), *1737* (SING); TENOMPOK: 1500 m, *Clemens 28641* (K, SING); ULAR HILL TRAIL: 1700 m, *Parris 11453* (K), 1800 m, *Parris & Croxall 9077* (K).

6.1.11. Asplenium filipes Copel., Philipp. J. Sci. 3C: 34 (1908). C. Chr. & Holttum, Gardens' Bull. 7: 278 (1934).

Terrestrial and lithophytic. Hill forest and lower montane forest, by streams. Elevation: 800–2100 m.

Material examined: DALLAS: 900 m, *Clemens 27292* (K, US); DALLAS/TENOMPOK: 1100 m, *Clemens 26894* (K, US); LUMU-LUMU: 2100 m, *Clemens 29969* (K); SUMMIT TRAIL: 1800 m, *Parris & Croxall 8882* (K).

6.1.12. Asplenium horizontale Baker, Ann. Bot. 8: 125 (1894). C. Chr. & Holttum, Gardens' Bull. 7: 277 (1934).

Lower montane forest.

Material examined: TENOMPOK: *Clemens 29266* (SING).

6.1.13. Asplenium kinabaluense Holttum in C. Chr. & Holttum, Gardens' Bull. 7: 281, pl. 62 (1934). Type: Victoria Peak: 4000 m, *Clemens 29830* (holotype SING!; isotype K!).

Terrestrial. *Rhododendron/Leptospermum* scrub above tree line. Elevation: 3300–4000 m. Endemic to Mount Kinabalu.

Additional material examined: GURULAU SPUR: 3700 m, *Clemens 50857* (K); LOW'S PEAK: 4000 m, *Clemens 27046b* (SING); PANAR LABAN: 3300 m, *Parris 11592* (K); SAYAT-SAYAT: 3700 m, *Parris & Croxall 8769* (K); SUMMIT TRAIL: 3500 m, *Parris & Croxall 8771* (K).

6.1.14. Asplenium klossii C. Chr. in C. Chr. & Holttum, Gardens' Bull. 7: 278, pl. 60 (1934).

Epiphyte. Lowland dipterocarp forest. Elevation: 500 m.

Material examined: PORING HOT SPRINGS: 500 m, *Parris & Croxall 9018* (K).

6.1.15. Asplenium lobangense C. Chr. in C. Chr. & Holttum, Gardens' Bull. 7: 279 (1934). Type: Lubang: 1200 m, *Holttum SFN 25548* (holotype BM n.v.; isotype SING!).

Terrestrial. Lower montane forest. Elevation: 1200 m. Not endemic to Mount Kinabalu.

6.1.16. Asplenium lobulatum Mett. ex Kuhn, Linnaea 36: 100 (1869).

Asplenium acutiusculum sensu C. Chr. & Holttum non Blume, Gardens' Bull. 7: 280 (1934).

Epiphytic, lithophytic and terrestrial. Lower montane oak forest in valleys. Elevation: 1100–2000 m.

Material examined: BAMBANGAN RIVER: 1500 m, *RSNB 1295* (SING); BUNDU TUHAN: 1200 m, *Clemens 28052* (US); DALLAS: 1100 m, *Clemens 26889* (US); KIAU VIEW TRAIL: 1600 m, *Parris & Croxall 8542* (K), 1600 m, *8543* (K); KIAU/LUBANG: *Topping 1814* (SING, US); KILEMBUN BASIN: 1400 m, *Clemens 33680* (US); LIWAGU RIVER TRAIL: 1500 m, *Parris 10823* (K), 1600 m, *11481* (K), 1600 m, *11482* (K), 1500 m, *11489* (K), 1400 m, *Parris & Croxall 8600* (K), 1400 m, *8601* (K), 1700 m, *9112* (K), 1700 m, *9113* (K), 1700 m, *9114* (K); MARAI PARAI: 1400 m, *Clemens 32381* (K), 1500 m, *33140* (K); MESILAU RIVER: 1500 m, *RSNB 4120* (K); PINOSUK PLATEAU: 1500 m, *Parris & Croxall 9144* (K), 1500 m, *9153* (K); SUMMIT TRAIL: 2000 m, *Collenette 21655* (US), 1700 m, *Parris & Croxall 8887* (K); TENOMPOK: 1500 m, *Clemens 26646* (K), *26889* (US), 1500 m, *27604* (K), 1500 m, *28054* (US), 1400 m, *Holttum SFN 25402* (K, SING).

6.1.17. Asplenium macrophyllum Sw., Schrad. J. Bot. 1800 (2): 52 (1801).

Epiphyte. Lowland dipterocarp forest. Elevation: 500 m.

Material examined: PORING HOT SPRINGS: 500 m, *Parris & Croxall 8986* (K).

6.1.18. Asplenium musifolium Mett., Aspl. 86, no. 3 (1859).

Epiphyte. Lower montane oak forest in valleys. Elevation: 1700–1900 m.

Material examined: MESILAU CAVE TRAIL: 1700–1900 m, *Beaman 9105* (K, MICH, MSC).

6.1.19. Asplenium nidus L., Sp. Pl., 1079 (1753). C. Chr. & Holttum, Gardens' Bull. 7: 275 (1934).

a. var. nidus

Asplenium nidus L. var. *ellipticum* (Fée) ined.?, C. Chr. & Holttum, Gardens' Bull. 7: 275 (1934).

Epiphyte. Lower montane forest. Elevation: 1200–1500 m.

Material examined: GURULAU SPUR: 1500 m, *Clemens 50482* (K); LUBANG: *Clemens 10324* (K); MAMUT ROAD: 1200–1300 m, *Tamura & Hotta 309* (K).

b. var. curtisorum (C. Chr.) Holttum, Gardens' Bull. 27: 148 (1974).

Epiphyte. Lowland dipterocarp forest. Elevation: 600 m.

Material examined: PORING HOT SPRINGS/LANGANAN WATER FALLS: 600 m, *Parris & Croxall 8950* (K).

6.1.20. Asplenium nigrescens Blume, Enum. Pl. Javae, 180 (1828). C. Chr. & Holttum, Gardens' Bull. 7: 280 (1934).

Terrestrial. Lower montane forest. Elevation: 1500 m.

Material examined: MARAI PARAI SPUR: 1500 m, *Holttum SFN 25615* (K, SING); MESILAU RIVER: 1500 m, *RSNB 1367* (SING); UPPER KINABALU: *Clemens 29972* (K).

6.1.21. Asplenium normale D. Don, Prodr. Fl. Nepal, 7 (1825). C. Chr. & Holttum, Gardens' Bull. 7: 278 (1934).

Terrestrial. Hill dipterocarp and lower montane oak forest near streams. Elevation: 800–1800 m.

Material examined: DALLAS: 1100 m, *Clemens 26890* (US); KIAU VIEW TRAIL: 1600 m, *Parris & Croxall 8541* (K); KINASARABAN HILL: 1400 m, *Sinclair et al. 8972* (K); LIWAGU RIVER TRAIL: 1700 m, *Parris 11476* (K); MAMUT RIVER: 1200 m, *RSNB 1737* (K, SING); MESILAU RIVER: 1800 m, *Clemens 29056* (K); SAYAP: 800–1000 m, *Beaman 9808* (K, MICH, MSC); SUMMIT TRAIL: 1800 m, *Parris & Croxall 8869* (K); TENOMPOK: 1500 m, *Clemens 26890* (K), 1500 m, *28165* (K), 1500 m, *28546* (US), 1500 m, *28879* (K, US), 1400 m, *Holttum SFN 25423* (K, SING); WEST MESILAU RIVER: 1600–1700 m, *Beaman 7456* (MICH, MSC).

6.1.22. Asplenium pachychlamys C. Chr. in C. Chr. & Holttum, Gardens' Bull. 7: 277 (1934). Type: Tenompok: 1400 m, *Holttum SFN 25355* (holotype BM n.v.; isotypes K!, SING!).

Epiphyte. Lower montane oak forest. Elevation: 1400–3000 m. Not endemic to Mount Kinabalu.

Additional material examined: GOLF COURSE SITE: 1700–1800 m, *Beaman 10663* (MICH, MSC); LIWAGU RIVER TRAIL: 1400 m, *Parris & Croxall 9035* (K), 1500 m, *9110* (K); LUBANG: *Topping 1777* (SING, US); MARAI PARAI: 1500 m, *Clemens 33029* (K); MENTEKI RIVER: 1600 m, *Beaman 10784* (K, MICH, MSC); PAKA-PAKA CAVE: 3000 m, *Clemens 27963* (SING); TENOMPOK: 1500 m, *Clemens 29228* (K); WEST MESILAU RIVER: 1600–1700 m, *Beaman 8664* (K, MICH, MSC), 1600 m, *9004* (MICH, MSC).

6.1.23. Asplenium paradoxum Blume, Enum. Pl. Javae, 179 (1828).

Lithophytic. Lowland dipterocarp forest. Elevation: 600 m.

Material examined: PORING HOT SPRINGS/LANGANAN WATER FALLS: 600 m, *Parris & Croxall 8920* (K).

6.1.24. Asplenium pellucidum Lam., Encycl. 2: 305 (1786).

Epiphyte. Lowland dipterocarp forest. Elevation: 500–600 m.

Material examined: NALUMAD: 600 m, *Shea & Aban SAN 77282* (K); PORING HOT SPRINGS: 500 m, *Parris & Croxall 9012* (K).

6.1.25. Asplenium persicifolium J. Sm. ex Mett., Aspl. 97, no. 32 (1859).

Lower montane forest.

Material examined: LUBANG: *Topping 1796* (US).

6.1.26. Asplenium phyllitidis D. Don

a. subsp. **malesicum** Holttum, Gardens' Bull. 27: 153 (1974).

Asplenium nidus L. var. *phyllitidis* (D. Don) Alderw., Handb. Malayan Ferns Suppl., 282 (1917). C. Chr. & Holttum, Gardens' Bull. 7: 275 (1934).

Epiphyte. Hill forest. Elevation: 900 m.

Material examined: DALLAS: 900 m, *Clemens 27304* (K).

6.1.27. Asplenium polyodon G. Forster, Prodr., 80 (1786).

Asplenium adiantoides (L.) C. Chr., Index Fil., 99 (1905) non Lam. (1786) nec Raddi (1819) nec Raoul (1844). C. Chr. & Holttum, Gardens' Bull. 7: 280 (1934).

Based on *Topping 1571* (BM? n.v.).

6.1.28. Asplenium scolopendrioides J. Sm. ex Hook., Icones Plantarum t. 930 (1854). C. Chr. & Holttum, Gardens' Bull. 7: 275 (1934).

Epiphytic, lithophytic, sometimes scandent. Hill forest and lower montane forest, near streams in deep shade. Elevation: 900–1500 m.

Material examined: DALLAS: 900 m, *Clemens 27373* (K); KIAU/TAHUBANG RIVER: 1000 m, *Parris 11532* (K); LIWAGU/MESILAU RIVERS: 1400 m, *RSNB 2897* (K, SING); PENATARAN BASIN: 1400 m, *Clemens 32591* (K); PENIBUKAN: 1200–1500 m, *Clemens 31469* (K); TAHUBANG RIVER: 900 m, *Low s.n.* (K); TENOMPOK: 1500 m, *Clemens 26830* (K), 1500 m, *27477* (K), 1500 m, *28457* (K), 1400 m, *Holttum SFN 25383* (K, SING).

6.1.29. Asplenium setisectum Blume, Enum. Pl. Javae, 187 (1828).

Asplenium malayo-alpinum Holttum in C. Chr. & Holttum, Gardens' Bull. 280, pl. 61 (1934). Type: Mount Kinabalu: 3800 m, *Holttum SFN 25485* (holotype SING!; isotype K!).

Lithophytic. Upper montane *Leptospermum, Dacrydium* and *Phyllocladus* forest. Elevation: 2900–3800 m.

Additional material examined. PAKA-PAKA CAVE: 3000 m, *Parris & Croxall 8785* (K); PANAR LABAN: 3300 m, *Parris 11596* (K); SUMMIT TRAIL: 3300 m, *Parris & Croxall 8770* (K), 3300 m, *8774* (K), 2900 m, *8794* (K).

6.1.30. Asplenium subnormale Copel., Perkins, Fragm., 183, t. 4, f. B (1905). C. Chr. & Holttum, Gardens' Bull. 7: 279 (1934).

Terrestrial. Lowland and hill dipterocarp forest. Elevation: 500–1000 m.

Material examined: DALLAS: 900 m, *Clemens 27298* (K); KIPUNGIT FALLS: 500 m, *Parris & Croxall 8916* (K); PORING HOT SPRINGS/LANGANAN WATER FALLS: 1000 m, *Parris & Croxall 8982* (K).

6.1.31. Asplenium tenerum G. Forster, Prodr., 80 (1786). C. Chr. & Holttum, Gardens' Bull. 7: 277 (1934).

Asplenium tenerum G. Forster var. *retusum* C. Chr. in C. Chr. & Holttum, Gardens' Bull. 7: 277 (1934). Type: Dallas/Tenompok: 1200–1500 m, *Clemens 27746* (syntype BM n.v.; isosyntype K!); Minitinduk: 900 m, *Holttum SFN 25586* (syntype BM n.v.; isosyntype SING!).

Epiphyte. Lower montane forest. Elevation: 900–2700 m.

Additional material examined: KEMBURONGOH: 2100 m, *Holttum SFN 25537* (K, SING); KILEMBUN RIVER: 2700 m, *Clemens 33827* (K); KINATEKI RIVER: 1200 m, *Clemens 31084* (K); KINATEKI RIVER HEAD: 2100–2700 m, *Clemens 33827* (K); LIWAGU RIVER TRAIL: 1400 m, *Parris & Croxall 8597* (K); LUMU-LUMU: 2100 m, *Clemens 29970* (K, SING); MAMUT RIVER: 1400 m, *RSNB 1721* (K, SING); MESILAU RIVER: 1400 m, *RSNB 1354* (K, SING); SUMMIT TRAIL: 2000 m, *Parris & Croxall 8691* (K).

6.1.32. Asplenium thunbergii Kunze, Linnaea 10: 517 (1836).

Asplenium belangeri Kunze, Bot. Zeit. 1848: 176 (1848). C. Chr. & Holttum, Gardens' Bull. 7: 278 (1934).

Epiphyte. Hill forest and lower montane forest. Elevation: 900–1400 m.

Material examined: DALLAS: 900 m, *Clemens 30378* (K); LIWAGU RIVER TRAIL: 1400 m, *Parris & Croxall 8596* (K); PENIBUKAN: 1200 m, *Clemens 32067* (K).

6.1.33. Asplenium trifoliatum Copel., Philipp. J. Sci. 5C: 284 (1910). C. Chr. & Holttum, Gardens' Bull. 7: 276 (1934).

Lithophyte. Lowland dipterocarp and lower montane forest. Elevation: 500–1700 m.

Material examined: DALLAS: 900 m, *Clemens 26818b* (SING); KIPUNGIT FALLS: 500 m, *Parris & Croxall 8906* (K); TENOMPOK: 1700 m, *Clemens 29429* (SING).

6.1.34. Asplenium unilaterale Lam., Encycl. 2: 305 (1786). C. Chr. & Holttum, Gardens' Bull. 7: 278 (1934).

Lithophyte. Lowland and hill dipterocarp forest, lower montane oak forest, on wet rocks in deep shade. Elevation: 500–2100 m.

Material examined: BAT CAVE: 600 m, *Parris & Croxall 8981* (K); DALLAS: 1100 m, *Holttum SFN 25146* (US); GURULAU SPUR: 1500–2100 m, *Clemens 50733* (K); KIAU: *Topping 1545* (US); KIAU/LUBANG: *Topping 1607* (US); KILEMBUN RIVER HEAD: 1400 m, *Clemens 32460* (K); KIPUNGIT FALLS: 500 m, *Parris & Croxall 8905* (K); LIWAGU RIVER TRAIL: 1500 m, *Parris 11495* (K), 1400 m, *Parris & Croxall 8598* (K), 1400 m, *8599* (K); LUBANG: *Topping 1793* (SING, US); LUGAS HILL: 1300 m, *Beaman 10529* (K, MICH, MSC); PENIBUKAN: 1200 m, *Clemens 30615* (K); SUMMIT TRAIL: 1800 m, *Parris & Croxall 8881* (K).

6.1.35. Asplenium vittiforme Cav., Descr., 255 (1802).

Asplenium squamulatum Blume, Enum. Pl. Javae, 174 (1828). C.Chr. & Holttum, Gardens' Bull. 7: 276 (1934).

Epiphyte. Lowland dipterocarp forest and hill forest. Elevation: 400–900 m.

Material examined: DALLAS: 900 m, *Clemens 27231* (K), 900 m, *27309* (US); KAUNG: 400 m, *Holttum SFN 25122* (SING), *Topping 1896* (SING); PORING HOT SPRINGS: 500 m, *Parris & Croxall 8984* (K).

6.1.36. Asplenium vulcanicum Blume, Enum. Pl. Javae, 176 (1828).

Asplenium filiceps Copel., Philipp. J. Sci. 5C: 285 (1910). C. Chr. & Holttum, Gardens' Bull. 7: 276 (1934).

Epiphytic and lithophytic. Lowland dipterocarp and hill forest. Elevation: 300–800 m.

Material examined: BAT CAVE: 600 m, *Parris & Croxall 8952* (K); DALLAS: 800 m, *Holttum SFN 25360* (K, SING); KEBAYAU: 300 m, *Clemens 27688* (SING); KIPUNGIT FALLS: 600 m, *Parris & Croxall 8907* (K); PORING HOT SPRINGS: 500 m, *Parris & Croxall 9017* (K).

6.1.37. Asplenium yoshinagae Makino

a. var. **planicaule** C. Morton, Contr. U. S. Natl. Herb. 38: 227 (1973).

Asplenium planicaule Wall. ex Mett., Abhandl. Senckenb. Naturf. Gessell. 3: 201 (1859) non Lowe (1858). C. Chr. & Holttum, Gardens' Bull. 7: 280 (1934).

Epiphyte. Lower montane forest. Elevation: 1500 m.

Material examined: LUBANG: *Topping 1773* (MICH); PINOSUK PLATEAU: 1500 m, *Parris & Croxall 9151* (K).

7. BLECHNACEAE

7.1. BLECHNUM

Copeland, E. B. 1960. Fern Flora of the Philippines. Vol. 3: 421–426. Holttum, R. E. 1968. Revised Flora of Malaya. Vol. 2, Ferns. Ed. 2: 444–449.

7.1.1. Blechnum borneense C. Chr., Dansk Bot. Arkiv 9: 56 (1937).

Terrestrial. Lower montane oak forest. Elevation: 1400–2600 m.

Material examined: EASTERN SHOULDER: 1500 m, *RSNB 1581* (K, SING); KILEMBUN RIVER: 1400 m, *Clemens 32512* (K); MESILAU: 1500 m, *RSNB 1341* (K); MESILAU RIVER: 1500 m, *RSNB 1341* (SING); SUMMIT TRAIL: 2600 m, *Parris & Croxall 8817* (K); ULAR HILL TRAIL: *Parris 11455* (K).

7.1.2. Blechnum egregium Copel., Perkins, Fragm., 187 (1905).

Terrestrial. Hill dipterocarp forest and lower montane forest. Elevation: 1000–1500 m.

Material examined: PENIBUKAN: 1200 m, *Clemens 30737* (K, SING), 1500 m, *31435* (K), 1200 m, *40812* (K); PINOSUK PLATEAU: 1400 m, *Beaman 10729* (MICH, MSC); PORING HOT SPRINGS/LANGANAN WATER FALLS: 1000 m, *Parris & Croxall 8927* (K).

7.1.3. Blechnum finlaysonianum Wall. ex Hook. & Grev., Ic. Fil., t. 225 (1831). C. Chr. & Holttum, Gardens' Bull. 7: 284 (1934).

Terrestrial. Lower montane forest. Elevation: 1500–2100 m.

Material examined: EASTERN SHOULDER: 1500 m, *RSNB 1582* (K, SING); KEMBURONGOH: 2100 m, *Holttum s.n.* (SING); TENOMPOK: 1500 m, *Clemens 28728* (K); TENOMPOK/TOMIS: 1700 m, *Clemens 29492* (K).

7.1.4. Blechnum fluviatile (R. Br.) Lowe ex Salomon, Nom., 115 (1883). C. Chr. & Holttum, Gardens' Bull. 7: 283 (1934).

Terrestrial. Upper montane forest. Elevation: 2700–4100 m. Disjunct from New Guinea, Australia, Tasmania and New Zealand.

Material examined: EASTERN SHOULDER: 3000 m, *RSNB 884* (K, SING); GURULAU SPUR: 3400 m, *Clemens 51450* (K); KEMBURONGOH/PAKA-PAKA CAVE: 2700–3700 m, *Clemens 28727* (K, SING); LOW'S PEAK: 4100 m, *Clemens 27047* (K); MARAI PARAI SPUR: 3000 m, *Clemens 33122* (US); PAKA-PAKA CAVE: 3500 m, *Molesworth-Allen 3283* (US), 3100 m, *Parris 11579* (K), 3000 m, *Parris & Croxall 8703* (K); SHEILA'S PLATEAU/SHANGRI LA VALLEY: 3400 m, *Collenette 21516* (K); SUMMIT AREA: 3300 m, *Sinclair et al. 9151* (K); SUMMIT TRAIL: 3300 m, *Sinclair et al. 9151* (SING).

7.1.5. Blechnum fraseri (Cunn.) Luerss., Flora 1876: 292 (1876).

Blechnum fraseri (Cunn.) Luerss. var. *philippinense* (Christ) C. Chr. in C. Chr. & Holttum, Gardens' Bull. 7: 284 (1934).

Terrestrial. Lower montane forest. Elevation: 1700–2700 m.

Material examined: GURULAU SPUR: 2400 m, *Clemens 51187* (K); KINATEKI RIVER HEAD: 2700 m, *Clemens 31957* (K); LIWAGU RIVER HEAD: 2300 m, *Meijer SAN 24138* (K); LIWAGU RIVER TRAIL: 1700 m, *Parris 11470* (K); MARAI PARAI SPUR: *Clemens 11032* (MICH); SUMMIT TRAIL: 2600 m, *Parris & Croxall 8811* (K).

7.1.6. Blechnum orientale L., Sp. Pl., 1077 (1753). C. Chr. & Holttum, Gardens' Bull. 7: 284 (1934).

Terrestrial. Hill forest. Elevation: 900 m.

Material examined: DALLAS: 900 m, *Holttum s.n.* (SING).

7.1.7. Blechnum pallescens T. C. Chambers, ined.

Terrestrial. Lower montane forest. Elevation: 2300 m. Not endemic to Mount Kinabalu.

Material examined: KILEMBUN RIVER: 2300 m, *Clemens 33725* (K, L).

7.1.8. Blechnum patersonii (R. Br.) Mett., Fil. Lips., 64, t. 4, f. 4–10 (1856). C. Chr. & Holttum, Gardens' Bull. 7: 283 (1934).

Terrestrial, lithophyte. Lower montane forest. Elevation: 2100 m. These specimens may not be correctly determined.

Material examined: KEMBURONGOH: 2100 m, *Holttum SFN 25528* (SING); KILEMBUN RIVER: *Clemens 34012* (MICH); KINATEKI RIVER HEAD: *Clemens 32417* (MICH).

7.1.9. Blechnum vestitum (Blume) Kuhn, Ann. Lugd. Bat. 4: 284 (1869). C. Chr. & Holttum, Gardens' Bull. 7: 284 (1934).

Terrestrial. Lower montane forest on ultramafic substrates. Elevation: 2100–2700 m.

Material examined: KEMBURONGOH: 2100 m, *Price 207* (K), 2100 m, *Sinclair et al. 9040* (K, SING); SUMMIT TRAIL: 2700 m, *Parris 11566* (K).

7.1.10. Blechnum vulcanicum (Blume) Kuhn, Ann. Lugd. Bat. 4: 284 (1869). C. Chr. & Holttum, Gardens' Bull. 7: 284 (1934).

Terrestrial. Lower montane forest, probably on ultramafic substrates. Elevation: 2600–2700 m.

Material examined: KILEMBUN RIVER HEAD: 2700 m, *Clemens 51760* (K); SUMMIT TRAIL: 2600 m, *Parris & Croxall 8800* (K).

7.1.11. Blechnum sp. 1

Blechnum procerum sensu C. Chr. & Holttum non (G. Forster) Sw., Gardens' Bull. 7: 283 (1934).

Terrestrial. Lower montane forest on landslides of ultramafic substrate. Elevation: 1900–2700 m. Endemic to Mount Kinabalu?

Material examined: KINATEKI RIVER HEAD: 2400–2700 m, *Clemens 31725* (K), 2100 m, *31773* (MICH); MESILAU CAVE: 1900–2200 m, *Beaman 9539* (K, MICH, MSC).

7.1.12. Blechnum sp. 2

Terrestrial. Lower montane forest. Elevation: 1500–2400 m. Endemic to Mount Kinabalu?

Material examined: GURULAU SPUR: 2100 m, *Clemens 50753* (K); KEMBURONGOH: 2400 m, *Clemens 28883* (K); TENOMPOK/LUMU-LUMU: 1500 m, *Holttum SFN 25443* (K, SING); UPPER KINABALU: *Clemens 27755* (K).

7.2. STENOCHLAENA

Holttum, R. E. 1968. Revised Flora of Malaya. Vol. 2, Ferns. Ed. 2: 410–413.

7.2.1. Stenochlaena palustris (N. L. Burm.) Bedd., Ferns Br. Ind. Suppl., 26 (1876).

Terrestrial climber. Lowland dipterocarp forest. Sight record of B. S. Parris, Poring Hot Springs, Nov. 1980.

8. CYATHEACEAE

Holttum, R. E. 1963. Cyatheaceae. Fl. Males. 2, 1 (2): 65–158.

8.1. CYATHEA

Holttum, R. E. 1974. The tree-ferns of the genus *Cyathea* in Borneo. Gardens' Bull. 27: 167–182.

8.1.1. Cyathea acanthophora Holttum, Kew Bull. 16: 51 (1962). Type: Mount Kinabalu: *Clemens 34012* (holotype BO n.v.; isotype US n.v.).

Tree fern. Lower montane forest. Elevation: 1800–2000 m. Endemic to Mount Kinabalu.

Additional material examined: SUMMIT TRAIL: 1800–2000 m, *Jacobs 5788* (K).

8.1.2. Cyathea brachyphylla Holttum, Gardens' Bull. 27: 181 (1974). Type: Pakapaka Cave: 3000 m, *E. F. Allen s.n.* (holotype K!).

Tree fern. Upper montane forest. Elevation: 3000 m. Type from a plant cultivated at Kew; endemic to Mount Kinabalu.

8.1.3. Cyathea capitata Copel., Philipp. J. Sci. 12C: 49 (1917). C. Chr. & Holttum, Gardens' Bull. 7: 218 (1934). Type: Marai Parai Spur: *Clemens 11033* (lectotype of Holttum, Fl. Males. 2, 1: 142 (1963), MICH!; isotype K!).

Tree fern. Lower montane forest. Elevation: 1500–2100 m. Also known from Mt. Murud.

Additional material examined: EASTERN SHOULDER: 2000 m, *RSNB 170* (K, SING); KIAU VIEW TRAIL: 1600 m, *Parris 10787* (K); LUMU-LUMU: 2000 m, *Clemens 27959* (K), 1800 m, *29109* (K), 2100 m, *29987* (K), 1700 m, *Holttum SFN 25447* (K, SING); MARAI PARAI: 1500 m, *Clemens 32298* (K), 1500 m, *32410* (K); MESILAU CAVE: 1800 m, *RSNB 4716* (K); PENATARAN BASIN: 1800 m, *Clemens 40142* (K); SUMMIT TRAIL: 1800–2000 m, *Jacobs 5790* (K), 1800 m, *Parris & Croxall 8878* (K); WEST MESILAU RIVER: 1600–1700 m, *Beaman 8625* (K, MICH, MSC).

8.1.4. Cyathea contaminans (Wall. ex Hook.) Copel., Philipp. J. Sci. 4C: 60 (1909). C. Chr. & Holttum, Gardens' Bull. 7: 222 (1934).

Cyathea contaminans (Wall. ex Hook.) Copel. var. *setulosa* (Hasskarl) C. Chr. in C. Chr. & Holttum, Gardens' Bull. 7: 222 (1934).

Tree fern. Lowland dipterocarp forest to lower montane forest. Elevation: 500–1600 m.

Material examined: BAMBANGAN RIVER: 1500–1600 m, *Hotta 20449* (K); DALLAS: 900 m, *Clemens 26861* (K), 1000 m, *Holttum s.n.* (SING); KIAU: *Topping 1518* (SING); KIAU/LUBANG: *Topping 1804* (SING); LUBANG: 1200 m, *Holttum SFN 25560* (K, SING); MINITINDUK GORGE: *Clemens 10446* (K); MOUNT KINABALU: 600 m, *Low s.n.* (K); NALUMAD: 500 m, *Shea & Aban SAN 77321* (K); PENATARAN RIVER: 900 m, *Clemens 32572* (K); PORING HOT SPRINGS: 500 m, *Parris & Croxall 9008* (K); TENOMPOK/KUNDASANG: 1500 m, *Sinclair et al. 9231* (K, SING).

8.1.5. Cyathea discophora Holttum, Kew Bull. 16: 54 (1962). Type: Marai Parai: 2400 m, *Clemens 31698* (holotype B n.v.; isotype K!).

Tree fern. In open place in lower montane forest on ultramafic substrate. Elevation: 2400 m. Known only from the type collection; endemic to Mount Kinabalu.

8.1.6. Cyathea glabra (Blume) Copel., Philipp. J. Sci. 4C: 35 (1909).

Cyathea vexans (Ces.) C. Chr. in C. Chr. & Holttum, Gardens' Bull. 7: 218 (1934).

Tree fern. Lower montane forest on ultramafic substrate.

Material examined: MARAI PARAI SPUR: *Clemens 10921* (K).

8.1.7. Cyathea havilandii Baker in Stapf, Trans. Linn. Soc. London, Bot. 4: 249 (1894). C. Chr. & Holttum, Gardens' Bull. 7: 221 (1934). Type: Mount Kinabalu: 3200 m, *Haviland 1485* (holotype K!).

Cyathea paleacea Copel., Philipp. J. Sci. 12C: 53 (1917). Type: Lubang/Paka-paka Cave: *Clemens 10726* (holotype PNH?†; isotype MICH!). *Cyathea rigida* Copel., Philipp. J. Sci. 12C: 53 (1917). Type: Lubang/Paka-paka Cave: *Topping 1758* (holotype PNH?†; isotype MICH!).

Tree fern. *Leptospermum* scrub on ultramafic substrate. Elevation: 2400– 3200 m. Endemic to Mount Kinabalu.

Additional material examined: KEMBURONGOH: 2600 m, *Meijer SAN 28578* (K); LAYANG-LAYANG/PAKA CAVE: 3000 m, *Edwards 2188* (K); PAKA-PAKA CAVE: 3000 m, *Clemens 27956* (K), 2700 m, *Holttum SFN 25509* (K, SING), 2400–2700 m, *Meijer SAN 22061* (K), 2900 m, *Sinclair et al. 9193* (K, SING); SUMMIT TRAIL: 2600 m, *Parris 11562* (K), 2900 m, *Parris & Croxall 8790* (K); UPPER KINABALU: *Clemens 29805* (K).

8.1.8. Cyathea latebrosa (Wall. ex Hook.) Copel., Philipp. J. Sci. 4C: 52 (1909). C. Chr. & Holttum, Gardens' Bull. 7: 222 (1934).

Tree fern. Lowland dipterocarp forest to lower montane forest. Elevation: 600–1600 m.

Material examined: DALLAS: 900 m, *Clemens 28158* (K), 800 m, *Holttum SFN 25625* (K, SING); EASTERN SHOULDER: 1100 m, *RSNB 203* (K, SING), 700 m, *1505* (K, SING); KIAU VIEW TRAIL: 1600 m, *Parris 10807* (K); LIWAGU RIVER HEAD: 1200 m, *RSNB 2854* (SING); LIWAGU RIVER TRAIL: 1500 m, *Parris 10827* (K); LIWAGU/MESILAU RIVERS: 1200 m, *RSNB 2854* (K); PORING HOT SPRINGS: 600 m, *Parris & Croxall 9000* (K); PORING HOT SPRINGS/LANGANAN WATER FALLS: 900 m, *Parris & Croxall 8957* (K); TENOMPOK: *Holttum SFN 25407* (K, SING).

8.1.9. Cyathea loheri Christ, Bull. Herb. Boissier II, 6: 1007 (1906).

Cyathea korthalsii sensu C. Chr. & Holttum non Mett., Gardens' Bull. 7: 222 (1934).

Tree fern. Lower montane and upper montane forest. Elevation: 1800–3000 m.

Material examined: EASTERN SHOULDER: 2900 m, *RSNB 790* (K, SING), 2700 m, *949a* (K); GOKING'S VALLEY: 2700 m, *Fuchs 21479* (K); KEMBURONGOH: 2100 m, *Holttum SFN 25527* (K, SING), 1800 m, *Meijer SAN 29113* (K); LAYANG-LAYANG: 2700 m, *Edwards 2187* (K), 2500 m, *2198* (K); MESILAU BASIN: 2400 m, *Clemens 29712* (K, SING); PAKA-PAKA CAVE: 3000 m, *Parris & Croxall 8787* (K), 3000 m, *Sinclair et al. 9192* (K, SING); SUMMIT TRAIL: 2500 m, *Parris & Croxall 8804* (K), 2600 m, *8842* (K).

8.1.10. Cyathea longipes Copel., Philipp. J. Sci. 12C: 54 (1917). C. Chr. & Holttum, Gardens' Bull. 7: 222 (1934). Type: Marai Parai Spur: *Clemens 10915* (lectotype of Holttum, Fl. Males. 2, 1: 98 (1963), MICH!; isolectotype K!).

Tree fern. Hill dipterocarp and lower montane forest, mostly on ultramafic substrates. Elevation: 1200–1700 m. Endemic to Mount Kinabalu.

Additional material examined: LIWAGU RIVER HEAD: 1400 m, *RSNB 2979* (SING); LIWAGU/MESILAU RIVERS: 1400 m, *RSNB 2979* (K); MARAI PARAI: 1500 m, *Clemens 33154* (K); MARAI PARAI SPUR: *Clemens 10915* (MICH), *Topping 1850* (SING); PENIBUKAN: 1500 m, *Clemens 30735* (K), 1200 m, *30963* (K, SING), 1200 m, *51757* (K), 1200 m, *Holttum SFN 25601* (K, SING), 1300 m, *Parris 11522* (K); PINOSUK PLATEAU: 1700 m, *RSNB 1886* (K, SING).

8.1.11. Cyathea megalosora Copel., Philipp. J. Sci. 12C: 54 (1917). C. Chr. & Holttum, Gardens' Bull. 7: 221 (1934). Type: Paka-paka Cave: *Topping 1759* (lectotype of Holttum, Fl. Males. 2, 1: 148 (1963), US n.v.; isolectotypes K!, SING!).

Tree fern. Lower montane mossy ridge forest. Elevation: 1800–3000 m. Not endemic to Mount Kinabalu.

Additional material examined: KEMBURONGOH/PAKA-PAKA CAVE: 2100–3000 m, *Gibbs 4237* (K); LAYANG-LAYANG: 2600 m, *Edwards 2196* (K); LUBANG/PAKA-PAKA CAVE: *Topping 1959* (SING); MESILAU BASIN: 2400 m, *Clemens 29068* (K, SING), 1800 m, *29714* (SING); MESILAU TRAIL: 2400 m, *Meijer SAN 38594* (K); SUMMIT TRAIL: 2500–2900 m, *Jacobs 5717* (K), 2200 m, *5773* (K).

8.1.12. Cyathea moluccana R. Br. in Desv., Mem. Soc. Linn. Paris 6: 322 (1827).

Cyathea kinabaluensis Copel., Philipp. J. Sci. 12C: 51 (1917). C. Chr. & Holttum, Gardens' Bull. 7: 218 (1934). Type: Gurulau Spur: *Clemens 10840* (holotype PNH?†).

Tree fern. Lowland dipterocarp forest to lower montane forest. Elevation: 800–1500 m.

Material examined: DALLAS/TENOMPOK: 1100 m, *Clemens 26824* (K), 1100 m, *27331* (SING); GURULAU SPUR: 1100–1200 m, *Clemens 51245* (K); PORING HOT SPRINGS: 800 m, *Meijer SAN 24042* (K); TENOMPOK: 1500 m, *Clemens 27331* (K).

8.1.13. Cyathea oosora Holttum, Kew Bull. 16: 59 (1962). Type: Gurulau Spur: 2400 m, *Clemens 51188* (holotype K!).

Tree fern. Lower montane to upper montane ridge forest. Elevation: 2200–3200 m. Known also from Sulawesi.

Additional material examined: LAYANG-LAYANG: 2600 m, *Parris 11599* (K); MARAI PARAI: 3200 m, *Clemens 33283* (K); SUMMIT TRAIL: 2200–2500 m, *Jacobs 5778* (K), 2600 m, *Parris & Croxall 8814* (K).

8.1.14. Cyathea polypoda Baker in Stapf, Trans. Linn. Soc. London, Bot. 4: 250 (1894). C. Chr. & Holttum, Gardens' Bull. 7: 219 (1934). Type: Mount Kinabalu: 2700 m, *Haviland 1479* (holotype K!).

Cyathea kemberangana Copel., Philipp. J. Sci. 12C: 52 (1917). Type: Kemburongoh: *Clemens 10500* (holotype PNH?†; isotype MICH!).

Tree fern. Lower montane ridge forest. Elevation: 1500–2700 m. Not endemic to Mount Kinabalu.

Additional material examined: KEMBURONGOH: 2300 m, *Edwards 2199* (K), 2100 m, *Meijer SAN 29150* (K); KEMBURONGOH/LUMU-LUMU: 1800 m, *Holttum SFN 25458* (K, SING); LIWAGU RIVER TRAIL: 1800 m, *Parris 11467* (K); LUMU-LUMU: 2100 m, *Clemens 27954* (K); SUMMIT TRAIL: 1800–2000 m, *Jacobs 5789* (K), 2100 m, *Parris & Croxall 8847* (K); TENOMPOK: 1500 m, *Clemens s.n.* (SING).

8.1.15. Cyathea ramispina (Hook.) Copel., Philipp. J. Sci. 4C: 36 (1909). C. Chr. & Holttum, Gardens' Bull. 7: 220 (1934).

Tree fern. Lower montane ridge forest. Elevation: 1200–2700 m.

Material examined: EASTERN SHOULDER: 1500 m, *RSNB 990* (K, SING); KEMBURONGOH/LUMU-LUMU: 1800 m, *Holttum SFN 25457* (K, SING); KIAU VIEW TRAIL: 1600 m, *Parris 10808* (K); KINATEKI RIVER: 1200–1500 m, *Clemens 31390* (SING); KINATEKI RIVER HEAD: 2700 m, *Clemens 31923* (SING); PENATARAN RIVER: 2400 m, *Clemens 32547* (K); SUMMIT TRAIL: 1800–2000 m, *Jacobs 5787* (K); TENOMPOK: 1500 m, *Clemens 28850* (K), 1500 m, *29491* (K, SING).

8.1.16. Cyathea recommutata Copel., Philipp. J. Sci. 4C: 36 (1909). C. Chr. & Holttum, Gardens' Bull. 7: 220 (1934).

Cyathea toppingii Copel., Philipp. J. Sci. 12C: 51 (1917). Type: Gurulau Spur: *Topping 1824* (holotype PNH?†; isotype SING!).

Tree fern. Lower montane oak forest. Elevation: 1200–1700 m.

Additional material examined: EASTERN SHOULDER: 1200 m, *RSNB 1526* (K, SING); GOLF COURSE SITE: 1700 m, *Beaman 8746* (K, MICH, MSC); KINASARABAN HILL: 1500 m, *Sinclair et al. 8983* (K, SING); MESILAU RIVER: 1500 m, *RSNB 4331* (K); TENOMPOK: 1500 m, *Clemens 28266* (K), 1500 m, *28596* (K), 1400 m, *Holttum SFN 25408* (K, SING).

8.1.17. Cyathea squamulata (Blume) Copel., Philipp. J. Sci. 4C: 37 (1909). C. Chr. & Holttum, Gardens' Bull. 7: 219 (1934).

Tree fern. Lower montane forest. Elevation: 1400–1800 m.

Material examined: SUMMIT TRAIL: 1400 m, *Jacobs 5021* (K), 1500–1800 m, *5701* (K); TENOMPOK/KUNDASANG: 1400 m, *Sinclair et al. 9234* (K, SING).

8.1.18. Cyathea stipitipinnula Holttum, Kew Bull. 16: 62 (1962). Type: Marai Parai: 1500 m, *Clemens 33156* (holotype K!).

Tree fern. Lower montane forest, apparently only on ultramafic substrates. Elevation: 1400–1500 m. Endemic to Mount Kinabalu.

Additional material examined: PENATARAN BASIN: 1400 m, *Clemens 34228* (K); PENIBUKAN: 1500 m, *Clemens 30906* (K, SING), 1500 m, *30962* (SING).

8.1.19. Cyathea trichodesma (Scort.) Copel., Philipp. J. Sci. 4C: 55 (1909).

Cyathea burbidgei sensu C. Chr. & Holttum p.p. non (Baker) Copel., Gardens' Bull. 7: 220 (1934).

Tree fern. Lower montane forest. Elevation: 1500 m.

Material examined: TENOMPOK: 1500 m, *Clemens 27741* (K), 1500 m, *29569* (K).

8.1.20. Cyathea trichophora Copel., Philipp. J. Sci. 6C: 363 (1911).

Cyathea elliptica Copel., Philipp. J. Sci. 12C: 51 (1917). Type: Gurulau Spur: *Clemens 10859* (lectotype of Holttum, Fl. Males. 2, 1: 146 (1963), MICH!). *Cyathea subbipinnata* Copel., Philipp. J. Sci. 56: 471, pl. 1 (1935). Type: Gurulau Spur: 1800 m, *Clemens 50396* (holotype PNH?†; isotype K!). *Cyathea holttumii* Copel., Philipp. J. Sci. 56: 472, pl. 2 (1935). Type: Tahubang River: 1100 m, *Clemens 40308* (holotype PNH?†; isotype K!).

Tree fern. Hill forest. Elevation: 1100–1800 m.

Additional material examined: GURULAU SPUR: *Clemens 10859* (MICH); KIAU VIEW TRAIL: 1600 m, *Parris 10789* (K); PENIBUKAN: 1100 m, *Clemens 40354* (K).

8.1.21. Cyathea tripinnata Copel., Philipp. J. Sci. 1, Suppl.: 251 (1906). C. Chr. & Holttum, Gardens' Bull. 7: 222 (1934).

Tree fern. Lower montane forest. Elevation: 1400–1500 m.

Material examined: KINASARABAN HILL: 1400 m, *Sinclair et al. 8984* (K, SING); LIWAGU/MESILAU RIVERS: 1400 m, *RSNB 2898* (K, SING); TENOMPOK: 1500 m, *Clemens 27937* (K), 1500 m, *28695* (K), 1500 m, *29395* (K, SING), 1400 m, *Holttum SFN 25409* (K, SING).

8.1.22. Cyathea wallacei (Mett. ex Kuhn) Copel., Philipp. J. Sci. 4C: 48 (1909).

Cyathea burbidgei (Baker) Copel., Philipp. J. Sci. 4C: 55 (1909). C. Chr. & Holttum, Gardens' Bull. 7: 220 (1934).

Tree fern. *Clemens 10839* and *Topping 1842* n.v., might be this species.

9. DAVALLIACEAE

9.1. ARAIOSTEGIA

9.1.1. Araiostegia hymenophylloides (Blume) Copel., Philipp. J. Sci. 34: 241 (1927).

Leucostegia hymenophylloides (Blume) Bedd., Ferns Brit. Ind. Suppl., 4 (1876). C.Chr. & Holttum, Gardens' Bull. 7: 231 (1934).

Epiphytic and terrestrial. Lower montane oak forest. Elevation: 1400–2100 m.

Material examined: KIAU VIEW TRAIL: 1600 m, *Parris 11614* (K), 1600 m, *Parris & Croxall 8540* (K); LUBANG: *Clemens 10388* (K); MENTEKI RIVER: 1600 m, *Beaman 10776* (K, MICH, MSC); MESILAU RIVER: 1500 m, *RSNB 1368* (K, SING), 2100 m, *Clemens 28446* (K); TENOMPOK: 1500–1800 m, *Clemens 27121* (K), 1500 m, *28097* (K), 1500 m, *28545* (K), 1400 m, *Holttum SFN 25353* (K, SING); WEST MESILAU RIVER: 1600–1700 m, *Beaman 8675* (MICH, MSC).

9.2. DAVALLIA

Copeland, E. B. 1958. Fern Flora of the Philippines. Vol. 1: 170–175. Holttum, R. E. 1968. Revised Flora of Malaya. Vol. 2, Ferns. Ed. 2: 354–363.

9.2.1. Davallia denticulata (N. L. Burm.) Mett. ex Kuhn, Fil. Deck., 27 (1867).

Epiphyte. Lowland dipterocarp forest. Elevation: 500 m.

Material examined: NALUMAD: 500 m, *Shea & Aban SAN 77266* (K); PORING HOT SPRINGS: 500 m, *Parris & Croxall 8988* (K).

9.2.2. Davallia embolostegia Copel., Philipp. J. Sci. 1, Suppl.: 147, t. 3 (1906). C. Chr. & Holttum, Gardens' Bull. 7: 233 (1934).

Epiphyte. Lower montane oak forest. Elevation: 1400–1600 m.

Material examined: KIAU VIEW TRAIL: 1600 m, *Parris 10802* (K); LUBANG: *Clemens 10356* (K), *Topping 1780* (SING); TENOMPOK: 1500 m, *Clemens 26145* (K), 1500 m, *27790* (K), 1500 m, *27803* (K), 1400 m, *Holttum SFN 25415* (K, SING).

9.2.3. Davallia lobbiana T. Moore, Index, 296 (1861). C. Chr. & Holttum, Gardens' Bull. 7: 233 (1934).

Epiphyte. Hill dipterocarp forest. Elevation: 1000 m.

Material examined: PORING HOT SPRINGS/LANGANAN WATER FALLS: 1000 m, *Parris & Croxall 8933* (K).

9.2.4. Davallia lorrainii Hance, Ann. Sci. Nat. V, 5: 254 (1866). C. Chr. & Holttum, Gardens' Bull. 7: 233 (1934).

Epiphyte. Elevation: 900–1200 m.

Material examined: KUNDASANG: 900–1200 m, *Clemens 29105* (K, SING).

9.2.5. Davallia solida (G. Forster) Sw., Schrad. J. Bot. 1800 (2): 87 (1801).

Epiphyte. Lowland dipterocarp forest. Sight record of B. S. Parris, Poring Hot Springs, Nov. 1980.

9.3. DAVALLODES

Holttum, R. E. 1972. The genus *Davallodes*. Kew Bull. 27: 245–249.

9.3.1. Davallodes borneense (Hook.) Copel., Philipp. J. Sci. 34: 250 (1927). C. Chr. & Holttum, Gardens' Bull. 7: 230 (1934).

Epiphyte. Lowland dipterocarp forest to lower montane forest. Elevation: 500–1500 m.

Material examined: KADAMAIAN RIVER NEAR MINITINDUK: 900 m, *Holttum SFN 25573* (K); KIAU/LUBANG: *Topping 1581* (SING); LUBANG: *Clemens 10322* (K); MINITINDUK: 900 m, *Holttum SFN 25573* (SING); PORING HOT SPRINGS: 500 m, *Parris & Croxall 9011* (K); TENOMPOK: 1500 m, *Clemens 29562* (K).

9.3.2. Davallodes burbidgei C. Chr. in C. Chr. & Holttum, Gardens' Bull. 7: 230 (1934). Type: Kadamaian River near Minitinduk: 900 m, *Holttum SFN 25567* (syntype BM n.v.; isosyntypes K!, SING!); Kaung: *Burbidge s.n.* (syntype K!).

Epiphyte. Hill forest, lower montane forest. Elevation: 900–1500 m. Not endemic to Mount Kinabalu.

Additional material examined: PENATARAN RIVER: 1200 m, *Clemens 34041* (K); PENIBUKAN: 1500 m, *Clemens 30954* (K), 1200–1500 m, *51699* (K), 1200 m, *51704* (K).

9.4. HUMATA

Holttum, R. E. 1968. Revised Flora of Malaya. Vol. 2, Ferns. Ed. 2: 364–371.

9.4.1. Humata heterophylla (Sm.) Desv., Prodr., 323 (1825).

Epiphyte. Lower montane dipterocarp forest. Elevation: 900 m.

Material examined: LOHAN/MAMUT COPPER MINE: 900 m, *Beaman 10624a* (MICH).

9.4.2. Humata repens (L. f.) Diels, Nat. Pfl. 1 (4): 209 (1899). C. Chr. & Holttum, Gardens' Bull. 7: 231 (1934).

Epiphyte. Hill dipterocarp forest, lower montane forest. Elevation: 800–2000 m.

Material examined: DALLAS: 800 m, *Holttum SFN 25725* (K, SING); GOLF COURSE SITE: 1700–2000 m, *Beaman 9590* (MICH, MSC); MINITINDUK: 900–1200 m, *Clemens 29626* (SING); PORING HOT SPRINGS/LANGANAN WATER FALLS: 1000 m, *Parris & Croxall 8967* (K); TENOMPOK: 1500 m, *Beaman 10522* (K, MICH, MSC), 1500 m, *Clemens 29626* (K).

9.4.3. Humata subvestita (C. Chr.) Parris, Fern Gaz. 12: 118 (1980).

Humata kinabaluensis Copel. var. *subvestita* C. Chr. in C. Chr. & Holttum, Gardens' Bull. 7: 232 (1934). Type: Lubang: 1400 m, *Holttum SFN 25549* (syntype BM n.v.; isosyntypes K!, SING!).

Epiphyte. Hill dipterocarp forest, lower montane oak forest. Elevation: 1000–3200 m. Not endemic to Mount Kinabalu.

Additional material examined: DALLAS/TENOMPOK: 1200 m, *Clemens 27007* (K); LIWAGU RIVER TRAIL: 1400 m, *Parris & Croxall 8591* (K), 1400 m, *8592* (K); MEMPENING TRAIL: 1600 m, *Parris 11498* (K); MOUNT KINABALU: 3200 m, *Haviland 1491* (K); PARK HEADQUARTERS: 1400 m, *Abbe et al. 9981* (K); PENATARAN BASIN: 1100 m, *Clemens 34197* (K, SING); PORING HOT SPRINGS/LANGANAN WATER FALLS: 1000 m, *Parris & Croxall 8930* (K); SUMMIT TRAIL: 2000 m, *Parris & Croxall 8888* (K); TENOMPOK: 1500 m, *Clemens 27816* (K), 1400 m, *Holttum SFN 25295* (K, SING).

9.4.4. Humata vestita (Blume) T. Moore, Index, 92 (1857).

Humata alpina (Blume) T. Moore, Index, 92 (1857). C. Chr. & Holttum, Gardens' Bull. 7: 231 (1934). *Humata kinabaluensis* Copel., Philipp. J. Sci. 12C: 48 (1917). C. Chr. & Holttum, Gardens' Bull. 7: 232 (1934). Type: Lubang/Paka-paka Cave: *Topping 1745* (holotype PNH?†; isotype MICH!).

Epiphyte. Lower montane and upper montane forest. Elevation: 1100–3200 m.

Additional material examined: EASTERN SHOULDER: 3200 m, *RSNB 774* (K, SING); EASTERN SHOULDER, CAMP 4: 2900 m, *RSNB 1158* (K); GURULAU SPUR: *Topping 1842* (MICH, SING); JANET'S HALT: 2400 m, *Collenette 563* (K); KADAMAIAN RIVER HEAD: 3200 m, *Clemens 50911* (K); KEMBURONGOH: 2400–3000 m, *Clemens 33181* (K), 2100 m, *Holttum SFN 25724* (K, SING); KINATEKI RIVER HEAD: 2700 m, *Clemens 31958* (SING); MARAI PARAI: 2400–3000 m, *Clemens 33181* (SING); MESILAU BASIN: 1800–2700 m, *Clemens 29063* (SING); MESILAU RIVER: 1800 m, *Clemens 51366* (K); MESILAU TRAIL: 2400–2600 m, *Meijer SAN 38593* (K); NUNGKEK LUBANG: 1100 m, *Clemens 32644* (SING); PAKA-PAKA CAVE: *Clemens 50911* (MICH), 3000 m, *Parris & Croxall 8791* (K); PENIBUKAN: 1200 m, *Clemens 32058* (SING); PINOSUK PLATEAU: 1200 m, *Cox 2549* (K); SUMMIT TRAIL: 2600 m, *Parris 11551* (K), 2600 m, *Parris & Croxall 8807* (K); UPPER KINABALU: *Clemens 29063* (K).

9.5. LEUCOSTEGIA

Holttum, R. E. 1968. Revised Flora of Malaya. Vol. 2, Ferns. Ed. 2: 351–353.

9.5.1. Leucostegia immersa C. Presl, Tent., 95, t. 4, f. 11 (1836). C. Chr. & Holttum, Gardens' Bull. 7: 231 (1934).

Terrestrial and epiphytic. Lower montane forest. Elevation: 1500–2100 m.

Material examined: LUMU-LUMU: 2100 m, *Clemens 29982* (K); MARAI PARAI: 1500 m, *Clemens 32415* (K); ULAR HILL TRAIL: 1800 m, *Parris & Croxall 9081* (K).

10. DENNSTAEDTIACEAE

10.1. DENNSTAEDTIA

Copeland, E. B. 1958. Fern Flora of the Philippines. Vol. 1: 88–95. Holttum, R. E. 1968. Revised Flora of Malaya. Vol. 2, Ferns. Ed. 2: 304–305.

10.1.1. Dennstaedtia ampla (Baker) Bedd., J. Bot. 31: 227 (1893). C. Chr. & Holttum, Gardens' Bull. 7: 227 (1934).

Terrestrial. Lower montane oak forest. Elevation: 1400–1500 m.

Material examined: LIWAGU RIVER TRAIL: 1500 m, *Parris 10826* (K); TENOMPOK: 1500 m, *Clemens 28281* (K), 1500 m, *28341* (K), 1500 m, *29435* (K), 1500 m, *29602* (K), 1400 m, *Holttum SFN 25406* (K, SING).

10.1.2. Dennstaedtia cuneata (J. Sm.) T. Moore, Index, 97 (1857). C. Chr. & Holttum, Gardens' Bull. 7: 227 (1934).

Terrestrial. Lower montane forest. Elevation: 1400–1500 m.

Material examined: KIAU: *Topping 1569* (SING); LUBANG: *Topping 1782* (SING); TENOMPOK: 1500 m, *Clemens 26976* (K), 1400 m, *Holttum SFN 25289* (K, SING).

10.1.3. Dennstaedtia rufidula C. Chr. in C. Chr. & Holttum, Gardens' Bull. 7: 226, pl. 51 (1934). Type: Dallas/Tenompok: 1200 m, *Holttum SFN 25298* (holotype BM n.v.; isotypes K!, SING!).

Terrestrial. Lower montane forest. Elevation: 1200–1500 m. Not endemic to Mount Kinabalu.

Additional material examined: MAMUT RIVER: 1400 m, *RSNB 1677* (K); TAHUBANG FALLS: 1200 m, *Clemens 40311* (K); TENOMPOK: 1500 m, *Clemens 26827* (K), 1500 m, *27465* (K).

10.1.4. Dennstaedtia scabra (Wall. ex Hook.) T. Moore

a. var. tenuisecta C. Chr. in C. Chr. & Holttum, Gardens' Bull. 7: 227 (1934). Type: Kemburongoh: 2100 m, *Holttum SFN 25520* (holotype BM n.v.; isotype SING!).

Terrestrial. Lower montane oak forest. Elevation: 1600–2100 m. Not endemic to Mount Kinabalu.

Additional material examined: LIWAGU RIVER TRAIL: 1600 m, *Parris & Croxall 9108* (K); ULAR HILL TRAIL: 1700 m, *Parris 11443* (K), 1800 m, *Parris & Croxall 9107* (K).

10.1.5. Dennstaedtia terminalis Alderw., Bull. Jard. Bot. Buit. 2, 16: 6, t. 4l (1914).

Terrestrial. Lower montane open ridge forest, sometimes on ultramafic substrates. Elevation: 1600–2700 m.

Material examined: KIAU VIEW TRAIL: 1600 m, *Parris 10806* (K); SUMMIT TRAIL: 2700 m, *Parris 11560* (K), 2600 m, *Parris & Croxall 8666* (K).

10.2. HISTIOPTERIS

Holttum, R. E. 1968. Revised Flora of Malaya. Vol. 2, Ferns. Ed. 2: 390–393.

10.2.1. Histiopteris incisa (Thunb.) J. Sm., Hist. Fil, 295 (1875). C. Chr. & Holttum, Gardens' Bull. 7: 289 (1934).

Terrestrial. Lower montane to upper montane forest in open situations. Elevation: 1400–3200 m.

Material examined: EASTERN SHOULDER: 2900 m, *RSNB 786* (K, SING); EASTERN SHOULDER, CAMP 4: 2900 m, *RSNB 1156* (K, SING); JANET'S HALT/SHEILA'S PLATEAU: 3200 m, *Collenette 21521* (K); MOUNT KINABALU: 2100 m, *Haviland 1492* (K); PAKA-PAKA CAVE: 3000 m, *Clemens 28018* (K), 3000 m, *Holttum SFN 25519* (K), 3100 m, *Parris 11582* (K), 3000 m, *Sinclair et al. 9190* (K, SING); TENOMPOK: 1400 m, *Holttum SFN 25414* (K, SING).

10.2.2. Histiopteris stipulacea (Hook.) Copel., Philipp. J. Sci. 3C: 347 (1909). C. Chr. & Holttum, Gardens' Bull. 7: 289 (1934).

Histiopteris stipulacea (Hook.) Copel. var. *integrifolia* (Copel.) C. Chr. in C. Chr. & Holttum, Gardens' Bull. 7: 289 (1934).

Terrestrial. Lower montane forest, in open situations. Elevation: 1200–1700 m.

Material examined: KIAU VIEW TRAIL: 1700 m, *Parris & Croxall 8550* (K); LIWAGU/MESILAU RIVERS: 1200 m, *RSNB 2731* (K, SING); PARK HEADQUARTERS: 1400 m, *Abbe et al. 9998* (K); TENOMPOK: 1500 m, *Clemens 28268* (K, SING), 1500 m, *28362* (K).

10.3. HYPOLEPIS

Brownsey, P. J. 1987. A review of the fern genus *Hypolepis* (Dennstaedtiaceae) in the Malesian and Pacific Regions. Blumea 32: 227–276.

10.3.1. Hypolepis alpina (Blume) Hook., Sp. Fil. 2: 63 (1852).

Terrestrial. Upper montane low ridge forest, in openings. Elevation: 1200–3200 m.

Material examined: EASTERN SHOULDER: 3000 m, *RSNB 923* (K, SING); PANAR LABAN: 3200 m, *Parris & Croxall 8781* (K); SUMMIT TRAIL: 3200 m, *Parris & Croxall 8781* (K); TAHUBANG FALLS: 1200 m, *Clemens 30702* (K).

10.3.2. Hypolepis bamleriana Rosenstock, Feddes Repert. 10: 325 (1912).

Hypolepis tenuifolia sensu C. Chr. & Holttum non (G. Forster) Bernh., Gardens' Bull. 7: 229 (1934).

Terrestrial. Lower montane oak forest. Elevation: 1600–2600 m.

Material examined: GURULAU SPUR: 1600 m, *Clemens 50708* (K); KEMBURONGOH: 2100 m, *Holttum SFN 25535* (K, SING); LIWAGU RIVER TRAIL: 1600 m, *Parris & Croxall 9109* (K); SUMMIT TRAIL: 2500 m, *Parris 11547* (K), 2600 m, *Parris & Croxall 8820* (K), 2400 m, *8868* (K).

10.3.3. Hypolepis brooksiae Alderw., Bull. Jard. Bot. Buit. 2, 28: 29 (1918). C. Chr. & Holttum, Gardens' Bull. 7: 229 (1934).

Terrestrial. Margins of lower montane forest. Elevation: 1400–1600 m.

Material examined: LIWAGU RIVER TRAIL: 1600 m, *Parris & Croxall 9038* (K); PAKA-PAKA CAVE: *Topping 1671* (SING); TENOMPOK: 1500 m, *Clemens 26809* (K), 1500 m, *27789* (K), 1500 m, *28366* (K), 1500 m, *29485* (K), 1400 m, *Holttum SFN 25411* (SING); TENOMPOK/KUNDASANG: 1500 m, *Sinclair et al. 9232* (SING).

10.3.4. Hypolepis glandulifera Brownsey & Chinn., J. Adelaide Bot. Gard. 10: 16 (1987).

Hypolepis punctata sensu C. Chr. & Holttum non (Thunb.) Mett. ex Kuhn, Gardens' Bull. 7: 229 (1934).

Terrestrial. Lower montane forest. Elevation: 900–1500 m.

Material examined: DALLAS/TENOMPOK: 1200 m, *Holttum SFN 25299* (K, SING); KILEMBUN BASIN: 1100 m, *Clemens 34397* (SING); KUNDASANG: 900 m, *Clemens 29110* (K), 900 m, *29239* (K); TENOMPOK: 1500 m, *Clemens 26980* (K).

10.3.5. Hypolepis pallida (Blume) Hook. ?, Sp. Fil. 2: 64 (1852).

Terrestrial. Elevation: 2700 m. Brownsey indicates that the specimen cited is atypical.

Material examined: EASTERN SHOULDER: 2700 m, *RSNB 952* (K, SING).

10.4. LINDSAEA

Kramer, K. U. 1971. *Lindsaea* Group. Fl. Males. 2, 1 (3): 198–254.

10.4.1. Lindsaea borneensis Hook. ex Baker, Syn. Fil. ed. 1, 107 (1867). C. Chr. & Holttum, Gardens' Bull. 7: 235 (1934).

Terrestrial. Lower montane forest. Elevation: 1200 m.

Material examined: MARAI PARAI SPUR: *Topping 1851* (SING, US); PENIBUKAN: 1200 m, *Clemens 30933* (K, SING).

10.4.2. Lindsaea bouillodii Christ, Not. Syst. 1: 59 (1909).

Lindsaea tenera sensu C. Chr. & Holttum non Dryander, Gardens' Bull. 7: 235 (1934).

Terrestrial. Hill dipterocarp forest, lower montane forest, by streams, sometimes on ultramafics. Elevation: 900–1500 m.

Material examined: DALLAS: 900 m, *Clemens 26895* (K), 1100 m, *Holttum SFN 25149* (K, SING); LIWAGU RIVER TRAIL: 1500 m, *Parris & Croxall 9052* (K); PINOSUK PLATEAU: 1400 m, *Beaman 10726* (MICH); PORING HOT SPRINGS/LANGANAN WATER FALLS: 1000 m, *Parris & Croxall 8964* (K).

10.4.3. Lindsaea carvifolia K. U. Kramer, Blumea 15: 569 (1968).

Epiphytic or terrestrial, scandent. Lower montane oak forest. Elevation: 1600–1900 m.

Material examined: MAMUT COPPER MINE: 1600–1700 m, *Beaman 9926* (K, MICH, MSC); MESILAU CAVE TRAIL: 1700–1900 m, *Beaman 7980* (MICH, MSC); WEST MESILAU RIVER: 1600 m, *Beaman 9028* (MICH, MSC).

10.4.4. Lindsaea crispa Baker, J. Bot. 17: 39 (1879). C. Chr. & Holttum, Gardens' Bull. 7: 238 (1934).

Lindsaea kinabaluensis Holttum in C. Chr. & Holttum, Gardens' Bull. 7: 237 (1934). Type: Tenompok/Lumu-lumu: 1500 m, *Holttum SFN 25433* (lectotype of Kramer, Fl. Males. 2, 1: 224 (1971), SING n.v.; isolectotype K!).

Terrestrial. Hill forest and lower montane forest. Elevation: 900–1800 m.

Additional material examined: DALLAS: 900 m, *Clemens 27744* (K, SING); EASTERN SHOULDER: 1500 m, *RSNB 1580b* (K); LIWAGU RIVER TRAIL: 1800 m, *Parris 11469* (K), 1500 m, *Parris & Croxall 9054* (K); LUBANG: 1200 m, *Holttum SFN 25728* (K, SING); TENOMPOK: 1500 m, *Clemens 28645* (K), 1400 m, *Holttum SFN 25420* (K, SING); ULAR HILL TRAIL: 1800 m, *Parris & Croxall 9076* (K).

10.4.5. Lindsaea cultrata (Willd.) Sw., Syn. Fil., 119 (1806).

Terrestrial. Hill forest. Elevation: 900 m.

Material examined: DALLAS: 900 m, *Clemens 27267* (K); KAUNG/DALLAS: 900 m, *Holttum SFN 25132* (K, SING).

10.4.6. Lindsaea gomphophylla Baker, Ann. Bot. 5: 204 (1891).

Terrestrial. Lower montane forest. Elevation: 1400 m.

Material examined: LIWAGU/MESILAU RIVERS: 1400 m, *RSNB 2975* (K, SING).

10.4.7. Lindsaea gueriniana (Gaud.) Desv., Mem. Soc. Linn. Paris 6: 312 (1827).

Schizoloma induratum (Baker) C. Chr., Index Fil., 618 (1906). C. Chr. & Holttum, Gardens' Bull. 7: 234 (1934).

Terrestrial. Low stature hill forest on ultramafic substrate. Elevation: 800–1000 m.

Material examined: LOHAN RIVER: 800–1000 m, *Beaman 9055* (K, MICH, MSC).

10.4.8. Lindsaea integra Holttum, Gardens' Bull. 5: 67, f. 6 (1930).

Lindsaea nitida sensu C. Chr. & Holttum non Copel., Gardens' Bull. 7: 237 (1934).

Terrestrial. Lower montane oak forest. Elevation: 1200–1700 m.

Material examined: KIAU: *Topping 1573* (US); KIAU VIEW TRAIL: 1600 m, *Parris & Croxall 8553* (K); LIWAGU RIVER TRAIL: 1500 m, *Parris & Croxall 9053* (K); MARAI PARAI SPUR: *Topping 1866* (SING, US); PENIBUKAN: 1200 m, *Holttum SFN 25602* (K, SING); PORING HOT SPRINGS/LANGANAN WATER FALLS: 1600 m, *Parris & Croxall 8936* (K); ULAR HILL TRAIL: 1700 m, *Parris 11438* (K).

10.4.9. Lindsaea jamesonioides Baker, J. Bot. 17: 39 (1879). Type: Mount Kinabalu: 2700 m, *Burbidge s.n.* (holotype K!; isotype BM n.v.).

Schizoloma jamesonioides (Baker) Copel., Philipp. J. Sci. 1, Suppl.: 252 (1906). C. Chr. & Holttum, Gardens' Bull. 7: 234 (1934).

Lithophytic. Lower montane *Dacrydium* forest, on ultrabasic substrates. Elevation: 1200–2700 m. Endemic to Mount Kinabalu.

Additional material examined: MARAI PARAI: *Clemens 10909* (K), *11063* (K), 1500 m, *32328* (US), 1700 m, *Haviland 1495* (K), 1500 m, *Holttum SFN 25607* (K); MARAI PARAI SPUR: 1500 m, *Holttum SFN 25607* (SING), 1600 m, *Parris 11527* (K), *Topping 1890* (SING, US, US); PENIBUKAN: 1200 m, *Clemens 30733* (K).

10.4.10. Lindsaea lobata Poir. in Lam., Encycl. Suppl. 3: 448 (1813).

Terrestrial. Lower montane forest. Elevation: 1400–1500 m.

Material examined: LIWAGU RIVER TRAIL: 1400 m, *Parris & Croxall 8593* (K), 1500 m, *9055* (K).

10.4.11. Lindsaea lucida Blume, Enum. Pl. Javae, 216 (1828).

a. subsp. **lucida**

Lindsaea concinna J. Sm. ex Hook., Sp. Fil. 1: 205, pl. 61B (1846). C. Chr. & Holttum, Gardens' Bull. 7: 235 (1934).

Presumably terrestrial. Probably hill forest.

Material examined: KAUNG/KIAU: *Topping 1504* (SING, US).

10.4.12. Lindsaea oblanceolata Alderw., Bull. Jard. Bot. Buit. 2, 23: 15 (1916).

Lindsaea pectinata sensu C. Chr. & Holttum non Blume, Gardens' Bull. 7: 235 (1934).

Terrestrial and scandent. Lower montane oak forest. Elevation: 1400–1700 m.

Material examined: GURULAU SPUR: *Topping 1820* (SING, US); KIAU VIEW TRAIL: 1600 m, *Parris & Croxall 8555* (K), 1600 m, A28556 (K); LIWAGU RIVER TRAIL: 1400 m, *Parris & Croxall 9058* (K); MARAI PARAI SPUR: 1500 m, *Holttum SFN 25608* (K, SING), *Topping 1871* (US); PARK HEADQUARTERS: 1500 m, *Parris 10786* (K); POWER STATION: 1700 m, *Parris & Croxall 9078* (K); TENOMPOK: 1500 m, *RSNB 1451* (K, SING), 1500 m, *Clemens 28597* (K), 1500 m, *Holttum SFN 25435* (SING); TENOMPOK/LUMU-LUMU: 1500 m, *Holttum SFN 25435* (K).

10.4.13. Lindsaea obtusa J. Sm. ex Hook., Sp. Fil. 1: 224 (1846).

Lindsaea decomposita sensu C. Chr. & Holttum non Willd., Gardens' Bull. 7: 236 (1934). *Lindsaea recurva* (Hook.) Hook., Sp. Fil. 1: 220, pl. 70A (1846). C. Chr. & Holttum, Gardens' Bull. 7: 236 (1934). *Lindsaea davallioides* sensu C. Chr. & Holttum non Blume, Gardens' Bull. 7: 238 (1934).

Terrestrial. Hill forest and lower montane forest. Elevation: 900–2300 m.

Material examined: KEMBURONGOH: 2100 m, *Holttum SFN 25729* (K, SING); LIWAGU RIVER TRAIL: 1400 m, *Parris & Croxall 9056* (K); PARK HEADQUARTERS: 1400 m, *Edwards 2165* (K), 1400 m, *2182* (K), 1400 m, *2183* (K); PORING HOT SPRINGS/LANGANAN WATER FALLS: 900 m, *Parris & Croxall 8963* (K); SUMMIT TRAIL: 1800 m, *Parris & Croxall 8883* (K), 2300 m, *9177* (K); TENOMPOK: 1500 m, *Clemens 28215* (US), 1400 m, *Holttum SFN 25403* (K, SING).

10.4.14. Lindsaea odorata Roxb., Calcutta J. Nat. Hist. 4: 511 (1844).

Lindsaea plumula Ridley, J. Malay. Br. Roy. As. Soc. 4: 22 (1926). C. Chr. & Holttum, Gardens' Bull. 7: 234 (1934).

Terrestrial. Lower montane forest, often in open situations. Elevation: 1000–2900 m.

Material examined: KILEMBUN RIVER: 2300 m, *Clemens 33724* (K), 2900 m, *33740* (K); LANGANAN FALLS: 1000 m, *Beaman 10991* (MSC); LIWAGU RIVER: 1400 m, *Parris & Croxall 8594* (K); MARAI PARAI: 1400 m, *Clemens 32810* (K); MESILAU CAVE: 1800 m, *RSNB 4710* (K); MESILAU RIVER: 1500 m, *RSNB 1374* (K, SING); PARK HEADQUARTERS: 1400 m, *Edwards 2160* (K); PARK HEADQUARTERS/POWER STATION: 1800 m, *Parris & Croxall 8890* (K); PENIBUKAN: 1200 m, *Clemens s.n.* (SING), 1700 m, *40292* (K); TENOMPOK: 1500 m, *Clemens 28603* (K, SING); ULAR HILL TRAIL: 1800 m, *Parris 11462* (K).

10.4.15. Lindsaea parallelogramma Alderw., Bull. Jard. Bot. Buit. 3, 5: 212 (1922). C. Chr. & Holttum, Gardens' Bull. 7: 238 (1934).

Terrestrial. Hill dipterocarp forest, lower montane forest. Elevation: 900–1500 m.

Material examined: EASTERN SHOULDER: 1500 m, *RSNB 1580a* (K, SING); LUGAS HILL: 1300 m, *Beaman 8439* (K, MICH, MSC); PORING HOT SPRINGS/LANGANAN WATER FALLS: 900 m, *Parris & Croxall 8961* (K), 900 m, *8962* (K); TAHUBANG RIVER: 900 m, *Holttum SFN 25597* (K, SING); TENOMPOK: 1500 m, *RSNB 1453* (K, SING).

10.4.16. Lindsaea parasitica (Roxb. ex Griffith) Hieron., Hedwigia 62: 14 (1920).

Lindsaea scandens Hook., Sp. Fil. 1: 205, pl. 63B (1846). C. Chr. & Holttum, Gardens' Bull. 7: 235 (1934).

The Burbidge specimen is unlocalized; *Topping 1871* (BM? n.v.).

10.4.17. Lindsaea repens (Bory) Thwaites

a. var. **pectinata** (Blume) Mett. ex Kuhn in Miq., Ann. Mus. Bot. Lugduno-Batavum 4: 277 (1869).

Lindsaea repens (Bory) Thwaites, C. Chr. & Holttum, Gardens' Bull. 7: 235 (1934) p.p.

Terrestrial/climber. Lower montane forest. Elevation: 1100–2100 m.

Material examined: DALLAS: 1100 m, *Holttum SFN 25145* (K, SING); KEMBURONGOH/LUMU-LUMU: 1800–2100 m, *Clemens 26651* (K); KINASARABAN HILL: 1500 m, *Sinclair et al. 8982* (K); PARK HEADQUARTERS: 1400 m, *Abbe et al. 9990* (K).

b. var. **pseudohemiptera** Alderw., Bull. Jard. Bot. Buit. 3, 2: 157, f. C (1920).

Epiphytic climber. Lower montane forest. Elevation: 1500–1700 m.

Material examined: KILEMBUN RIVER: 1700 m, *Clemens 34108* (US).

c. var. **sessilis** (Copel.) Kramer, Blumea 15: 568 (1968).

Lindsaea repens (Bory) Thwaites, C. Chr. & Holttum, Gardens' Bull. 7: 235 (1934) p.p.

Scandent epiphyte. Hill dipterocarp and lower montane forest. Elevation: 600–1700 m.

Material examined: BAMBANGAN RIVER: 1700 m, *Clemens 33055* (K); GURULAU SPUR: *Topping 1841* (MICH, SING, US); KIAU: *Clemens 10231* (MICH), *Topping 1536* (MICH, SING, US), *1567* (MICH); KIAU VIEW TRAIL: 1600 m, *Parris & Croxall 8554* (K); KIAU/LUBANG: *Topping 1807* (MICH, SING, US); KUNDASANG: 1400 m, *RSNB 1464* (K, SING); LIWAGU RIVER HEAD: 1400 m, *RSNB 2896* (SING); LIWAGU RIVER TRAIL: 1500 m, *Parris 10825* (K); LIWAGU/MESILAU RIVERS: 1400 m, *RSNB 2896* (K); LUGAS HILL: 1300 m, *Beaman 8121* (K, MICH, MSC); MAMUT RIVER: 1200 m, *RSNB 1742* (K, SING); MELANGKAP KAPA: 600–700 m, *Beaman 8586* (K, MICH, MSC); TENOMPOK: 1500 m, *Clemens 27189* (K), 1500 m, *28826* (K), 1500 m, *29507* (K).

10.4.18. Lindsaea rigida J. Sm. ex Hook., Sp. Fil. 1: 217, pl. 63A (1846).

Lindsaea diplosora Alderw., Bull. Jard. Bot. Buit. 2, 16: 21 (1914). C. Chr. & Holttum, Gardens' Bull. 7: 235 (1934). *Lindsaea diplosora* Alderw. var. *acrosora* C. Chr. in C. Chr. & Holttum, Gardens' Bull. 7: 235 (1934).

Terrestrial or epiphytic. Lower montane forest, usually on ridges. Elevation: 1200–2400 m.

Material examined: GURULAU SPUR: 1500 m, *Clemens 50593* (K); KEMBURONGOH: 2400 m, *Clemens 28963* (K); KEMBURONGOH/LUMU-LUMU: 1800 m, *Holttum SFN 25459* (K, SING); LIWAGU RIVER TRAIL: 1500 m, *Parris & Croxall 9092* (K); LUMU-LUMU: 1800 m, *Clemens 27061* (K); MT. TIBABAR: 1900 m, *Sinclair et al. 9032* (K, SING); PENIBUKAN: 1500 m, *Clemens 30903* (K), 1200 m, *31088* (K); PINOSUK PLATEAU: 1700 m, *RSNB 1811* (K, SING); SUMMIT TRAIL: 2100 m, *Parris 11612* (K), 2100 m, *Parris & Croxall 8889* (K); TENOMPOK: 1500 m, *Clemens s.n.* (SING), 1500 m, *28963b* (K); TENOMPOK/KEMBURONGOH: 2400 m, *Clemens 28001* (K).

10.5. MICROLEPIA

Copeland, E. B. 1958. Fern Flora of the Philippines. Vol. 1: 95–99. Holttum, R. E. 1968. Revised Flora of Malaya. Vol. 2, Ferns. Ed. 2: 306–316. Tagawa, M. & Iwatsuki, K. 1979. Flora of Thailand. Vol. 3, Part 1: 112–124.

10.5.1. Microlepia hookeriana (Wall. ex Hook.) C. Presl, Epim., 95 (1849). C. Chr. & Holttum, Gardens' Bull. 7: 228 (1934).

Nephrolepis (?) marginalis Copel., Philipp. J. Sci. 12C: 49 (1917). Type: Gurulau Spur: *Topping 1632* (holotype PNH?†; isotype MICH!).

Terrestrial. Hill forest and lower montane forest. Elevation: 900–1800 m.

Additional material examined: DALLAS: 900 m, *Clemens 27407* (K), 1100 m, *Holttum SFN 25138* (K, SING); PARK HEADQUARTERS: 1600 m, *Parris 11438* (K), 1500 m, *Parris & Croxall 8569* (K); PENIBUKAN: 1200 m, *Clemens 30490* (K, SING); TENOMPOK: 1500–1800 m, *Clemens 27950* (K), 1500 m, *29092* (K), 1500 m, *30460* (K).

10.5.2. Microlepia manilensis (Goldm.) C. Chr., Index Fil., 427 (1906). C. Chr. & Holttum, Gardens' Bull. 7: 228 (1934).

Terrestrial. Lower montane forest. Elevation: 1200–2100 m.

Material examined: DALLAS/TENOMPOK: 1400 m, *Clemens 27527* (K); KIAU VIEW TRAIL: 1600 m, *Parris 10788* (K); LIWAGU RIVER TRAIL: 1700 m, *Parris 11477* (K), 1200 m, *Parris & Croxall 9100* (K); LUMU-LUMU: 2100 m, *Clemens 29941* (SING); SUMMIT TRAIL: 1800 m, *Parris & Croxall 8866* (K); TENOMPOK: 1500 m, *Clemens 27527* (SING), 1500 m, *28175* (K, SING), 1500 m, *28236* (K, SING), 1500 m, *28570* (K), 1500 m, *29683* (SING), 1400 m, *Holttum SFN 25381* (K, SING); TENOMPOK/KUNDASANG: 1500 m, *Sinclair et al. 9229* (K, SING).

10.5.3. Microlepia ridleyi Copel., Philipp. J. Sci. 11C: 39 (1916). C. Chr. & Holttum, Gardens' Bull. 7: 228 (1934).

Terrestrial. Lower montane forest. Elevation: 1200–1500 m.

Material examined: TENOMPOK: 1500 m, *Clemens 26978* (K), 1200 m, *Holttum SFN 25296* (K, SING).

10.5.4. Microlepia speluncae (L.) T. Moore

a. var. **villosissima** C. Chr., Gardens' Bull. 4: 399 (1929). C. Chr. & Holttum, Gardens' Bull. 7: 229 (1934).

Terrestrial. Lowland and hill dipterocarp forest, open areas. Elevation: 600–900 m.

Material examined: DALLAS: 900 m, *Clemens 27672* (SING); KEBAYAU/KAUNG: *Topping 1907* (MICH); PORING HOT SPRINGS/LANGANAN WATER FALLS: 600 m, *Parris & Croxall 8951* (K).

10.5.5. Microlepia strigosa (Thunb.) C. Presl, Epim., 95 (1849). C. Chr. & Holttum, Gardens' Bull. 7: 228 (1934).

Terrestrial. Lower montane forest. Elevation: 1200–1500 m.

Material examined: LUBANG: 1200 m, *Holttum SFN 25564* (K, SING); TENOMPOK: 1500 m, *Clemens 27907* (K).

10.6. ORTHIOPTERIS

Holttum, R. E. 1968. Revised Flora of Malaya. Vol. 2, Ferns. Ed. 2: 305–306.

10.6.1. Orthiopteris kingii (Bedd.) Holttum, Rev. Fl. Malaya 2: 306, f. 175 (1955).

Terrestrial. Lower montane forest. Elevation: 1500 m.

Material examined: EASTERN SHOULDER: 1500 m, *RSNB 1578* (SING).

10.7. PAESIA

Copeland, E. B. 1949. Pteridaceae of New Guinea. Philipp. J. Sci. 78: 28–29. Copeland, E. B. 1958. Fern Flora of the Philippines. Vol. 1: 122–123.

10.7.1. Paesia radula (Baker) C. Chr., Index Fil., 476 (1906). C. Chr. & Holttum, Gardens' Bull. 7: 229 (1934).

Terrestrial. Lower to upper montane forest in open situations. Elevation: 2100–3400 m.

Material examined: GURULAU SPUR: 3400 m, *Clemens 50908* (K), 2400 m, *50930* (K); KEMBURONGOH: 2700 m, *Clemens 27957* (K); KEMBURONGOH/PAKA-PAKA CAVE: 2700 m, *Holttum SFN 25474* (K, SING); MARAI PARAI: 3200 m, *Clemens 33208* (K); MESILAU: 2100 m, *Clemens 29047* (K); SHANGRI LA VALLEY: 3400 m, *Collenette 21505* (K); SUMMIT TRAIL: 2400 m, *Parris 11545* (K), 3000 m, *Parris & Croxall 8786* (K), 2600 m, *8819* (K); UPPER KINABALU: *Clemens 28974* (K).

10.8. PTERIDIUM

Holttum, R. E. 1968. Revised Flora of Malaya. Vol. 2, Ferns. Ed. 2: 388–390.

10.8.1. Pteridium esculentum (G. Forster) Cockayne, Rep. Bot. Survey Tongariro Natl. Park, 34 (1908). C. Chr. & Holttum, Gardens' Bull. 7: 230 (1934).

Terrestrial. Hill forest and lower montane forest, in open areas. Elevation: 1000–2600 m.

Material examined: DALLAS: 1000 m, *Holttum SFN 25377* (K, SING); KEMBURONGOH: 2600 m, *Parris & Croxall 8843* (K); KIAU: *Topping 1515* (SING); KIAU VIEW TRAIL: 1600 m, *Parris 11621* (K); KUNDASANG: *Clemens 29107* (K).

10.9. SPHENOMERIS

Kramer, K. U. 1971. *Lindsaea* Group. Fl. Males. 2, 1 (3): 179–184.

10.9.1. Sphenomeris chinensis (L.) Maxon, J. Wash. Acad. Sci. 3: 144 (1913). C. Chr. & Holttum, Gardens' Bull. 7: 234 (1934).

a. var. chinensis

Sphenomeris chinensis (L.) Maxon, Chr. & Holttum, Gardens' Bull. 7: 234 (1934).

Terrestrial and lithophytic. Hill forest and lower montane forest, in open areas. Elevation: 900–2100 m.

Material examined: BAMBANGAN RIVER: 1500 m, *RSNB 1303* (K); DALLAS: 900 m, *Clemens 29161* (K), 900 m, *Holttum s.n.* (SING); KAUNG: *Burbidge s.n.* (K); MEMPENING TRAIL: 1600 m, *Parris 11509* (K); PENATARAN BASIN: 2100 m, *Clemens 34140* (K); SOSOPODON: 1400 m, *Shim SAN 81801* (K); SUMMIT TRAIL: 1800 m, *Parris & Croxall 8844* (K).

10.9.2. Sphenomeris retusa (Cav.) Maxon, J. Wash. Acad. Sci. 3: 144 (1913).

Terrestrial. Hill forest. Elevation: 1100 m.

Material examined: TAHUBANG RIVER: 1100 m, *Kanis SAN 51463* (K).

10.9.3. Sphenomeris veitchii (Baker) C. Chr. in C. Chr. & Holttum, Gardens' Bull. 7: 234 (1934).

Davallia veitchii Baker, J. Bot. 17: 39 (1879). Type: Mount Kinabalu: 1800–2100 m, *Burbidge s.n.* (holotype K!).

Terrestrial. Lower montane forest. Elevation: 1700–2100 m. Endemic to Mount Kinabalu.

Additional material examined: MESILAU CAVE: 1800 m, *RSNB 4785* (K, SING); MESILAU CAVE TRAIL: 1700–1900 m, *Beaman 7996* (K, MICH, MSC).

10.10. TAPEINIDIUM

Kramer, K. U. 1971. *Lindsaea* Group. Fl. Males. 2, 1 (3): 184–197.

10.10.1. Tapeinidium acuminatum K. U. Kramer, Blumea 15: 554 (1968).

Terrestrial. Hill forest on ultramafic substrate. Elevation: 900 m.

Material examined: HEMPUEN HILL: 900 m, *Meijer SAN 20989* (K).

10.10.2. Tapeinidium biserratum (Blume) Alderw., Handb. Malayan Ferns Suppl., 509 (1917). C. Chr. & Holttum, Gardens' Bull. 7: 233 (1934) p.p.?

Terrestrial. Lower montane forest on ultramafic substrate. Elevation: 1400 m.

Material examined: PENIBUKAN: 1400 m, *Parris 11528* (K).

10.10.3. Tapeinidium luzonicum (Hook.) K. U. Kramer, Blumea 15: 552 (1968).

a. var. **luzonicum**

Tapeinidium biserratum sensu C. Chr. & Holttum p.p. non (Blume) Alderw., Gardens' Bull. 7: 233 (1934).

Terrestrial. Hill forest and lower montane forest. Elevation: 900–2600 m.

Material examined: DALLAS: 900 m, *Clemens 27839* (K); EASTERN SHOULDER: 1200 m, *RSNB 1529* (K, SING); KILEMBUN RIVER: 2600 m, *Clemens 33696* (K); KILEMBUN RIVER HEAD: 1400 m, *Clemens 32461* (K); LUBANG: 1200 m, *Holttum SFN 25546* (K, SING); LUBANG/KEMBURONGOH: *Topping 1643* (MICH); MAMUT COPPER MINE: 1600–1700 m, *Beaman 9955* (K, MICH, MSC); MEMPENING TRAIL: 1600 m, *Parris 11505* (K); PINOSUK PLATEAU: 1400 m, *Beaman 10755* (K, MICH, MSC); PORING HOT SPRINGS: 1200 m, *Meijer SAN 24071* (K); PORING HOT SPRINGS/LANGANAN WATER FALLS: 1000 m, *Parris & Croxall 8939* (K), 900 m, *8959* (K); SINGH'S PLATEAU: 900 m, *RSNB 1022* (K, SING); TENOMPOK: 1500 m, *Clemens 28350* (K).

10.10.4. Tapeinidium pinnatum (Cav.) C. Chr., Index Fil., 631 (1906). C. Chr. & Holttum, Gardens' Bull. 7: 233 (1934).

Terrestrial. Hill forest and lower montane forest. Elevation: 800–1300 m.

Material examined: PENIBUKAN: 1200 m, *Holttum SFN 25603* (K, SING), 1300 m, *Parris 11523* (K); PORING HOT SPRINGS/LANGANAN WATER FALLS: 1000m, *Parris & Croxall 8938* (K); TAKUTAN: 800 m, *Shea & Aban SAN 77239* (K).

11. DICKSONIACEAE

11.1. CIBOTIUM

Holttum, R. E. 1963. Cyatheaceae. Fl. Males. 2, 1 (2): 164–166.

11.1.1. Cibotium arachnoideum (C. Chr.) Holttum, Fl. Males. II, 1: 166 (1963).

Cibotium cumingii Kunze var. *arachnoideum* C. Chr. in C. Chr. & Holttum, Gardens' Bull. 7: 224 (1934). Type: Dallas: 900 m, *Holttum SFN 25378* (lectotype of Holttum, Fl. Males. 2, 1: 166 (1963), K!; isolectotypes BM n.v., SING!).

Terrestrial with massive prostrate rhizome. Hill forest to lower montane forest, generally in secondary situations. Elevation: 700–2100 m. Not endemic to Mount Kinabalu.

Additional material examined: DALLAS: 900 m, *Clemens 26852* (K), 900 m, *27740* (K), 900 m, *29162* (K); EASTERN SHOULDER: 800 m, *RSNB 998* (K, SING); GURULAU SPUR: 1600 m, *Clemens 50709* (K), 1500 m, *Gibbs 3999* (K); KIAU: *Topping 1514* (SING), *1520* (SING); LUMU-LUMU: 2100 m, *Clemens 29672* (SING); MELANGKAP KAPA: 700–1000 m, *Beaman 8784* (K, MICH, MSC); PARK HEADQUARTERS: 1500 m, *Parris & Croxall 9059* (K); PENATARAN BASIN: 1500 m, *Clemens 34137* (K); PORING HOT SPRINGS: 900–1200 m, *Meijer SAN 24059* (K); TENOMPOK: 2100 m, *Clemens 29672* (K).

11.2. CULCITA

Holttum, R. E. 1963. Cyatheaceae. Fl. Males. 2, 1 (2): 166–169.

11.2.1. Culcita javanica (Blume) Maxon, J. Wash. Acad. Sci. 12: 456 (1922).

Culcita copelandii (Christ) Maxon, J. Wash. Acad. Sci. 12: 457 (1922). C. Chr. & Holttum, Gardens' Bull. 7: 224 (1934).

Tree fern or rhizome prostrate. Hill dipterocarp forest to lower montane forest. Elevation: 1000–1600 m.

Material examined: BUNDU TUHAN: 1400 m, *Sinclair et al. 8999* (K, SING); KIAU VIEW TRAIL: 1600 m, *Parris & Croxall 8548* (K); MEMPENING TRAIL: 1600 m, *Parris 11500* (K); PARK HEADQUARTERS: 1600 m, *Parris 11517* (K); PORING HOT SPRINGS/LANGANAN WATER FALLS: 1000 m, *Parris & Croxall 8923* (K); TENOMPOK: 1500 m, *Clemens 27742* (K, SING).

11.3. CYSTODIUM

Holttum, R. E. 1963. Cyatheaceae. Fl. Males. 2, 1 (2): 162–164.

11.3.1. Cystodium sorbifolium (Sm.) J. Sm. in Hook., Gen. Fil., t. 96 (1841). C. Chr. & Holttum, Gardens' Bull. 7: 239 (1934).

Tree fern or rhizome prostrate. Lowland dipterocarp forest. Elevation: 300–500 m.

Material examined: KEBAYAU: 300 m, *Clemens 27677* (K); KEBAYAU/KAUNG: 300 m, *Holttum SFN 25106* (K, SING); PORING HOT SPRINGS: 500 m, *Parris & Croxall 8991* (K).

11.4. DICKSONIA

Holttum, R. E. 1963. Cyatheaceae. Fl. Males. 2, 1 (2): 158–162.

11.4.1. Dicksonia mollis Holttum, Kew Bull. 16: 64 (1962).

Dicksonia blumei sensu C. Chr. & Holttum non (Kunze) T. Moore, Gardens' Bull. 7: 223 (1934).

Tree fern. Lower montane forest. Elevation: 1400–2400 m.

Material examined: KEMBURONGOH: 2000 m, *Meijer SAN 29188* (K, SING); KIAU VIEW TRAIL: 1600 m, *Parris & Croxall 8549* (K); KINATEKI RIVER HEAD: 2100 m, *Clemens 31743* (SING), 2100–2400 m, *31992* (K); LIWAGU/MESILAU RIVERS: 1400 m, *RSNB 2899* (K, SING); MARAI PARAI: 1500 m, *Clemens 29055A* (K); MESILAU CAVE: 1800 m, *RSNB 4715* (K, SING); MINIRINTEG (MESILAU BASIN): 2400 m, *Clemens 29055* (SING); MT. LENAU (BETWEEN TENOMPOK AND KEMBURONGOH): 1500 m, *Sinclair et al. 9005* (K, SING); TENOMPOK: 1500 m, *Clemens 28734* (SING), 1500 m, *29734* (K); TENOMPOK/LUMU-LUMU: 1500 m, *Holttum SFN 25445* (K, SING); UPPER KINABALU: *Clemens 29055* (K).

12. DIPTERIDACEAE

12.1. CHEIROPLEURIA

Holttum, R. E. 1968. Revised Flora of Malaya. Vol. 2, Ferns. Ed. 2: 136–137.

12.1.1. Cheiropleuria bicuspis (Blume) C. Presl, Epim., 189 (1849). C. Chr. & Holttum, Gardens' Bull. 7: 314 (1934).

Terrestrial. Lower montane forest, sometimes on ultramafic substrates. Elevation: 1200–2400 m.

Material examined: EASTERN SHOULDER: 1400 m, *RSNB 588* (K, SING); GURULAU SPUR: *Gibbs 3992* (K), 1700 m, *3998* (K), *Topping 1625* (SING); KEMBURONGOH: 2100 m, *Holttum 48* (SING); KEMBURONGOH/LUMU-LUMU: 1800 m, *Holttum SFN 25451* (K, SING); KILEMBUN RIVER HEAD: 1800 m, *Clemens 32466* (K); LIWAGU RIVER TRAIL: 1500 m, *Parris & Croxall 8610* (K); LUBANG/PAKA-PAKA CAVE: *Topping 1750* (SING); LUMU-LUMU: 2100 m, *Clemens 29334* (K); MAMUT COPPER MINE: 1400–1500 m, *Beaman 10330* (MICH, MSC); MEMPENING TRAIL: 1600 m, *Parris 11513* (K); MOUNT KINABALU: 1200–2400 m, *Burbidge s.n.* (K); TENOMPOK: 1500 m, *Clemens 28339* (K).

12.2. DIPTERIS

Holttum, R. E. 1968. Revised Flora of Malaya. Vol. 2, Ferns. Ed. 2: 132–136.

12.2.1. Dipteris conjugata Reinw., Syll. Pl. 2: 3 (1824). C. Chr. & Holttum, Gardens' Bull. 7: 314 (1934).

Terrestrial. Hill forest and lower montane forest, in open situations. Elevation: 700–2600 m.

Material examined: MELANGKAP KAPA: 700–1000 m, *Beaman 8787* (MICH, MSC); MESILAU CAVE: 1900–2200 m, *Beaman 9540* (MICH, MSC); SUMMIT TRAIL: 2600 m, *Parris & Croxall 8845* (K), 2600 m, *8846 (K)*.

12.2.2. Dipteris novoguineensis Posth., Rec. Trav. Bot. Neerl. 25a: 248, f. 1 (1928).

Terrestial. Upper levels of lower montane ridge forest and upper montane ridge forest. Elevation: 2400–2900 m.

Material examined: JANET'S HALT/SHEILA'S PLATEAU: 2900 m, *Collenette 21554* (K); LAYANG-LAYANG: 2600 m, *Parris 11601* (K); MOUNT KINABALU: 2400 m, *Holttum s.n.* (SING); SUMMIT TRAIL: 2500 m, *Parris & Croxall 8809* (K).

13. DRYOPTERIDACEAE

13.1. ACROPHORUS

Holttum, R. E. 1968. Revised Flora of Malaya. Vol. 2, Ferns. Ed. 2: 480–481.

13.1.1. Acrophorus nodosus C. Presl, Tent., 94, t. 3, f. 2 (1836).

Acrophorus blumei Ching. C.Chr. & Holttum, Gardens' Bull. 7: 226 (1934).

Terrestrial. Lower montane oak forest to low upper montane forest. Elevation: 1500–3700 m.

Material examined: EASTERN SHOULDER: 3000 m, *RSNB 735* (K, SING); EASTERN SHOULDER, CAMP 4: 2900 m, *RSNB 1155* (K), 2900 m, *1155b* (SING); GURULAU SPUR: 3700 m, *Clemens 50858* (K); LIWAGU RIVER HEAD: 2300 m, *Meijer SAN 24134* (K); MESILAU BASIN: 2100 m, *Clemens 29966* (SING); PAKA-PAKA CAVE: 3000 m, *Holttum SFN 25499* (K, SING); SUMMIT TRAIL: 1500–3000 m, *Clemens 27753* (K), 2300 m, *Parris 11540* (K), 3400 m, *Parris & Croxall 8780* (K); TENOMPOK: 1500 m, *Clemens 28837* (K); UPPER KINABALU: *Clemens 28977* (K), *29276* (K), *29966* (K).

13.2. AENIGMOPTERIS

Holttum, R. E. 1984. Studies in the fern-genera allied to *Tectaria*. III. *Aenigmopteris* and *Ataxipteris*, two new genera allied to *Tectaria* Cav., with comments on *Psomiocarpa* Presl. Blumea 30: 1–11.

13.2.1. Aenigmopteris elegans Holttum, Blumea 30: 8 (1984). Type: Park Headquarters: 1600 m, *Parris & Croxall 9135* (holotype K!).

Terrestrial. Lower montane forest. Elevation: 1200–1600 m. Endemic to Mount Kinabalu.

Additional material examined: GIGISSEN CREEK/MARAI PARAI: 1400 m, *Clemens 32386* (SING); LIWAGU RIVER TRAIL: 1400 m, *Parris & Croxall 8604* (K), 1400 m, *8612* (K); PENIBUKAN: 1200 m, *Clemens 30676* (MICH).

13.3. ARACHNIODES

Holttum, R. E. 1968. Revised Flora of Malaya. Vol. 2, Ferns. Ed. 2: 484–488. Sledge, W. A. 1973. The dryopteroid ferns of Ceylon. Bull. Brit. Mus. (Nat. Hist.) Bot. 5: 38–43.

13.3.1. Arachniodes aristata (G. Forster) Tind., Contrib. N. S. W. Nat. Herb. 3: 89 (1961).

Polystichum aristatum (G. Forster) C. Presl, Tent., 83 (1836). C. Chr. & Holttum, Gardens' Bull. 7: 258 (1934).

Terrestrial. Lower montane forest. Elevation: 1200 m.

Material examined: LUBANG: *Holttum SFN 25566* (K, SING).

13.3.2. Arachniodes puncticulata (Alderw.) Ching, Acta Bot. Sin. 10: 259 (1962).

Polystichum puncticulatum Alderw., Bull. Jard. Bot. Buit. 3, 2: 171 (1920). C. Chr. & Holttum, Gardens' Bull. 7: 257 (1934).

Terrestrial. Lowland to lower montane forest. Elevation: 300–2000 m.

Material examined: DALLAS/TENOMPOK: 1200 m, *Clemens 27328* (K); KEBAYAU/KAUNG: 300 m, *Clemens 27660* (K, SING); KEMBURONGOH/LUMU-LUMU: 2000 m, *Sinclair et al. 9205* (K, SING); MESILAU CAVE: 1800 m, *RSNB 4717* (K); TENOMPOK: 1500 m, *Clemens 28032* (K), 1500 m, *28102* (K), 1400 m, *Holttum SFN 25359* (K).

13.3.3. Arachniodes tripinnata (Goldm.) Sledge, Bull. Brit. Mus. (Nat. Hist.) Bot. 5: 41 (1973).

Polystichum carvifolium (Kunze) C. Chr., Index Fil., 70 (1905). C. Chr. & Holttum, Gardens' Bull. 7: 258 (1934).

Terrestrial. Lower montane oak forest. Elevation: 1700–2600 m.

Material examined: KADAMAIAN FALLS TRAIL: 1700 m, *Parris & Croxall 9075* (K); KEMBURONGOH: 2100 m, *Holttum SFN 25525* (K, SING); KILEMBUN BASIN: 2300–2600 m, *Clemens 33821* (K); KINATEKI/PENATARAN RIVERS: 2100–2300 m, *Clemens 31953* (K); LUMU-LUMU: 1800 m, *Clemens 27969* (K); SUMMIT TRAIL: 2200 m, *Parris 11536* (K), 2400 m, *Parris & Croxall 8867* (K); TENOMPOK/LUMU-LUMU: *Holttum SFN 25439* (K, SING); ULAR HILL TRAIL: 1700 m, *Parris 11442* (K).

13.4. CTENITIS

Holttum, R. E. 1985. Studies in the fern genera allied to *Tectaria* Cav. IV. The genus *Ctenitis* in Asia, Malesia and the western Pacific. Blumea 31: 1–38.

13.4.1. Ctenitis aciculata (Baker) Ching, Bull. Fan Mem. Inst. Biol. Bot. 8: 292 (1938).

Terrestrial. Hill dipterocarp forest. Elevation: 1000 m.

Material examined: PORING HOT SPRINGS/LANGANAN WATER FALLS: 1000 m, *Parris & Croxall 8940* (K), 1000 m, *8941* (K).

13.4.2. Ctenitis atrorubens Holttum, Blumea 31: 29 (1985).

Terrestrial. Lower montane forest on ultramafic substrate. Elevation: 1400 m. Known only from two collections, the other from Philippines.

Material examined: PINOSUK PLATEAU: 1400 m, *Beaman 10734* (K, MICH, MSC).

13.4.3. Ctenitis kinabaluensis Holttum, Blumea 31: 16 (1985).

a. var. **kinabaluensis** Type: Dallas: 1100 m, *Holttum SFN 25253* (holotype K!; isotype SING!).

Dryopteris aciculata sensu C. Chr. & Holttum non (Baker) C. Chr., Gardens' Bull. 7: 253 (1934).

Terrestrial. Hill dipterocarp and lower montane forest. Elevation: 1000–1500 m. Not endemic to Mount Kinabalu.

Additional material examined: EASTERN SHOULDER, CAMP 1: 1200 m, *RSNB 1212* (K, SING); PORING HOT SPRINGS/LANGANAN WATER FALLS: 1000 m, *Parris & Croxall 8942* (K); TENOMPOK: 1500 m, *Clemens 26846* (K).

13.4.4. Ctenitis minutiloba Holttum, Blumea 31: 23 (1985). Type: Minteleb Hill: 600 m, *Meijer SAN 20267* (holotype K!).

Terrestrial. Hill dipterocarp forest. Elevation: 600–1000 m. Endemic to Mount Kinabalu.

Additional material examined: PORING HOT SPRINGS/LANGANAN WATER FALLS: 1000 m, *Parris & Croxall 8925* (K).

13.5. CYCLOPELTIS

Holttum, R. E. 1968. Revised Flora of Malaya. Vol. 2, Ferns. Ed. 2: 526–527.

13.5.1. Cyclopeltis crenata (Fée) C. Chr., Index Fil. suppl. 3: 64 (1934).

Cyclopeltis presliana (J. Sm.) Berkeley, Introd. Crypt. Bot., 517 (1857). C. Chr. & Holttum, Gardens' Bull. 7: 258 (1934).

Terrestrial. Lowland forest. Elevation: 300 m.

Material examined: KEBAYAU/KAUNG: 300 m, *Holttum SFN 25626* (K, SING).

13.6. DIACALPE

13.6.1. Diacalpe aspidioides Blume, Enum. Pl. Javae, 241 (1828). C. Chr. & Holttum, Gardens' Bull. 7: 226 (1934).

Terrestrial. Lower and upper montane forest. Elevation: 1200–2700 m.

Material examined: EASTERN SHOULDER: 2700 m, *RSNB 947* (K, SING); KILEMBUN BASIN: 2400 m, *Clemens 32517* (K); KILEMBUN RIVER: 1500–2400 m, *Clemens 33684* (K); KINATEKI RIVER: 1200–1500 m, *Clemens 31396* (K); LUBANG: *Clemens 10355* (K); MESILAU BASIN: 2100 m, *Clemens 29046* (K, SING); PENIBUKAN: 2100 m, *Clemens 50029* (K); SUMMIT TRAIL: 2200 m, *Parris 11535* (K); WASAI RIVER (PENATARAN BASIN): 1200 m, *Clemens 34134* (K).

13.7. DIDYMOCHLAENA

Holttum, R. E. 1968. Revised Flora of Malaya. Vol. 2, Ferns. Ed. 2: 483–484.

13.7.1. Didymochlaena truncatula (Sw.) J. Sm., J. Bot. 4: 196 (1841). C. Chr. & Holttum, Gardens' Bull. 7: 254 (1934).

Terrestrial, almost tree-fern like. Lower montane forest. Elevation: 900–1200 m.

Material examined: DALLAS/TENOMPOK: 900–1200 m, *Clemens 30455* (K); EASTERN SHOULDER: 1100 m, *RSNB 667* (K, SING); MINITINDUK: 900 m, *Holttum s.n.* (SING); MOUNT KINABALU: 1100 m, *Burbidge s.n.* (K); PENIBUKAN: 1200 m, *Clemens 30668* (K), 1200 m, *40650* (K).

13.8. DRYOPSIS

Holttum, R. E. & Edwards, P. J. 1986. Studies in the fern-genera allied to *Tectaria*. II. *Dryopsis*, a new genus. Kew Bull. 41: 171–204.

13.8.1. Dryopsis adnata (Blume) Holttum & P. J. Edwards, Kew Bull. 41: 190 (1986).

Dryopteris adnata (Blume) Alderw., Handb. Ferns Malaya, 191 (1908). C. Chr. & Holttum, Gardens' Bull. 7: 252 (1934).

Terrestrial. Upper montane forest. Elevation: 3000–3700 m.

Material examined: MOUNT KINABALU: 3500 m, *Holttum SFN 25488* (K, SING); PAKA-PAKA CAVE: 3000 m, *Holttum SFN 25514* (SING), 3100 m, *Parris 11572* (K), 3100 m, *11574* (K), 3000 m, *Sinclair et al. 9183* (K, SING), 3000 m, *9183A* (SING); SUMMIT AREA: 3700 m, *Clemens 51539* (K); SUMMIT TRAIL: 3100 m, *Parris 11583* (K), 3300 m, *Parris & Croxall 8742* (K), 3300 m, *8766* (K); UPPER KINABALU: *Clemens 27966* (K), *28973* (K).

13.8.2. Dryopsis paucisora (Copel.) Holttum & P. J. Edwards, Kew Bull. 41: 183 (1986).

Terrestrial? Lower montane forest? Elevation: 1200 m.

Material examined: PENIBUKAN: 1200 m, *Clemens 30552* (SING).

13.9. DRYOPTERIS

Holttum, R. E. 1968. Revised Flora of Malaya. Vol. 2, Ferns. Ed. 2: 490–493. Price, M. G. 1977. Philippine *Dryopteris*. Gardens' Bull. 30: 239–250. Sledge, W. A. 1973. The dryopteroid ferns of Ceylon. Bull. Brit. Mus. (Nat. Hist.) Bot. 5: 5–27. Tagawa, M. & Iwatsuki, K. 1988. Flora of Thailand. Vol. 3, Part 3: 345–356.

13.9.1. Dryopteris chaerophyllifolia (Zipp.) C. Chr., Index Fil., 257 (1905).

Dryopteris subarborea sensu C. Chr. & Holttum p.p. non (Baker) C. Chr., Gardens' Bull. 7: 252 (1934).

Terrestrial. Lower montane oak forest. Elevation: 1800 m.

Material examined: SUMMIT TRAIL: 1800 m, *Parris & Croxall 8891* (K); ULAR HILL TRAIL: 1800 m, *Parris & Croxall 9088* (K); UPPER KINABALU: *Clemens 29710* (K).

13.9.2. Dryopteris hasseltii (Blume) C. Chr., Index Fil., 269 (1905). C. Chr. & Holttum, Gardens' Bull. 7: 254 (1934).

Terrestrial. Lower montane forest. Elevation: 1500 m.

Material examined: TENOMPOK: 1500 m, *Clemens 28598* (K).

13.9.3. Dryopteris polita Rosenstock, Feddes Rep. 13: 218 (1914). C. Chr. & Holttum, Gardens' Bull. 7: 252 (1934).

Terrestrial. Lower montane oak forest. Elevation: 1400–1500 m.

Material examined: LIWAGU RIVER TRAIL: 1500 m, *Parris 11491* (K), 1400 m, *Parris & Croxall 8609* (K); TENOMPOK: 1500 m, *Clemens 27976* (K, SING), 1400 m, *Holttum SFN 25387* (SING).

13.9.4. Dryopteris purpurascens (Blume) Christ, Philipp. J. Sci. 2C: 213 (1907).

Dryopteris subarborea sensu C. Chr. & Holttum p.p. non (Baker) C. Chr., Gardens' Bull. 7: 252 (1934).

Terrestrial. Lower montane forest. Elevation: 1400–2900 m.

Material examined: EASTERN SHOULDER: 2900 m, *RSNB 834* (K), 2900 m, *1155A* (K); KEMBURONGOH/LUMU-LUMU: 1400 m, *Holttum SFN 25456* (K, SING); KINATEKI RIVER HEAD: 2700 m, *Clemens 31955* (K); MARAI PARAI: 1800 m, *Clemens 32653* (K, SING); SUMMIT TRAIL: 2500 m, *Parris 11548* (K), 2500 m, *Parris & Croxall 8802* (K); UPPER KINABALU: *Clemens 29875* (K).

13.9.5. Dryopteris scottii (Bedd.) Ching, Bull. Dept. Biol. Sun Yatsen Univ. 6: 3 (1933).

Dryopteris hirtipes sensu C. Chr. & Holttum p.p. non (Blume) Kuntze, Gardens' Bull. 7: 250 (1934).

Terrestrial. Lower montane forest. Elevation: 1200–1500 m.

Material examined: KILEMBUN RIVER: 1400 m, *Clemens 33955* (K); LIWAGU RIVER TRAIL: 1400 m, *Parris & Croxall 9032* (K); LUBANG: 1200 m, *Holttum SFN 25557* (SING); PINOSUK

PLATEAU: 1500 m, *Parris & Croxall 9148* (K); TENOMPOK: 1500 m, *Clemens 28205* (K), 1500 m, *29416* (K), 1400 m, *Holttum SFN 25390* (SING).

13.9.6. Dryopteris sparsa (D. Don) Kuntze, Rev. Gen. Pl. 2: 813 (1891). C. Chr. & Holttum, Gardens' Bull. 7: 252 (1934).

Terrestrial. Lower montane forest. Elevation: 900–2100 m.

Material examined: KADAMAIAN RIVER NEAR MINITINDUK: 900 m, *Holttum SFN 25576* (K); LIWAGU RIVER TRAIL: 1500 m, *Parris 10822* (K), 1400 m, *Parris & Croxall 9036* (K); LUMU-LUMU: 2100 m, *Clemens 29940* (SING); MEMPENING TRAIL: 1600 m, *Parris 11516* (K); MESILAU BASIN: *Clemens 29057* (SING); PARK HEADQUARTERS: 1500 m, *Parris & Croxall 8570* (K); UPPER KINABALU: *Clemens 29057* (K).

13.9.7. Dryopteris subarborea (Baker) C. Chr., Index Fil., 295 (1905). C. Chr. & Holttum, Gardens' Bull. 7: 252 (1934) p.p.

Terrestrial. Elevation: 2700 m.

Material examined: KINATEKI RIVER HEAD: 2700 m, *Clemens 31955* (SING).

13.9.8. Dryopteris wallichiana (Spreng.) Hyl., Bot. Notiser 1953: 352 (1953).

Dryopteris paleacea (Sw.) C. Chr., Amer. Fern J. 1: 94 (1911). C. Chr. & Holttum, Gardens' Bull. 7: 250 (1934).

Terrestrial. Upper montane forest. Elevation: 2600–3100 m.

Material examined: EASTERN SHOULDER: 2900 m, *RSNB 918* (K, SING); GOKING'S VALLEY: 2800 m, *Fuchs 21480* (K); KILEMBUN RIVER: 2600 m, *Clemens 33753* (K); PAKA-PAKA CAVE: 3100 m, *Parris 11573* (K).

13.10. HETEROGONIUM

Holttum, R. E. 1975. The genus *Heterogonium* Presl. Kalikasan 4: 205–231.

13.10.1. Heterogonium aspidioides C. Presl, Epim., 143 (1849).

Heterogonium profereoides (Christ) Copel., Univ. Calif. Publ. Bot. 16: 61 (1929). C. Chr. & Holttum, Gardens' Bull. 7: 258 (1934).

Terrestrial. Lowland dipterocarp and hill forest. Elevation: 600–1100 m.

Material examined: DALLAS: 900 m, *Holttum SFN 25257* (K, SING); KIPUNGIT FALLS TRAIL: 600 m, *Parris & Croxall 8911* (K).

13.10.2. Heterogonium sagenioides (Mett.) Holttum, Sarawak Mus. J. 5: 161 (1949).

Terrestrial. Lowland dipterocarp forest. Elevation: 500 m.

Material examined: PORING HOT SPRINGS: 500 m, *Parris & Croxall 9002* (K).

13.11. NOTHOPERANEMA

Price, M. G. 1977. Philippine *Dryopteris*. Gardens' Bull. 30: 239–240.

13.11.1. Nothoperanema hendersonii (Bedd.) Ching, Acta Phytotax. Sin. 11: 28 (1966).

Terrestrial ? Elevation: 2600 m.

Material examined: KILEMBUN BASIN: 2600 m, *Clemens 33750* (K, SING).

13.11.2. Nothoperanema sp. 1

Terrestrial. Lower montane forest. Elevation: 2600 m. Endemic to Mount Kinabalu.

Material examined: SUMMIT TRAIL: 2600 m, *Parris & Croxall 8815* (K).

13.12. PLEOCNEMIA

Holttum, R. E. 1974. The fern-genus *Pleocnemia*. Kew Bull. 29: 341–357.

13.12.1. Pleocnemia hemiteliiformis (Racib.) Holttum, Reinwardtia 1: 179, f. 11 (1951).

Terrestrial. Lowland dipterocarp forest to lower montane forest. Elevation: 600–1500 m.

Material examined: EASTERN SHOULDER: 800 m, *RSNB 579* (K, SING); LIWAGU/MESILAU RIVERS: 1200 m, *RSNB 2853* (K, SING); PORING HOT SPRINGS: 600 m, *Parris & Croxall 8999* (K); TENOMPOK: 1500 m, *Clemens 28287* (K).

13.12.2. Pleocnemia irregularis (C. Presl) Holttum, Kew Bull. 29: 347 (1974).

Terrestrial. Lowland dipterocarp forest. Sight record of B. S. Parris, Kipungit Falls, Nov. 1980.

13.12.3. Pleocnemia olivacea (Copel.) Holttum, Reinwardtia 1: 181, f. 8 (1951).

Tectaria leuzeana sensu C. Chr. & Holttum non (Gaud.) Copel., Gardens' Bull. 7: 259 (1934).

Terrestrial. Lower montane forest. Elevation: 1200–1600 m.

Material examined: KIAU: *Clemens 10241* (K); KIAU VIEW TRAIL: 1600 m, *Parris 10796* (K); LUBANG: 1200 m, *Holttum SFN 25545* (SING); TENOMPOK: 1500 m, *Clemens 27578* (K), 1500 m, *28749* (K), 1500 m, *28852* (K), 1500 m, *29111* (K), 1500 m, *29629* (K).

13.13. POLYSTICHUM

Copeland, E. B. 1960. Fern Flora of the Philippines. Vol. 2: 242–248.

13.13.1. Polystichum gemmiparum C. Chr. in C. Chr. & Holttum, Gardens' Bull. 7: 256, pl. 54 (1934). Type: Lubang: 1200 m, *Holttum SFN 25561* (holotype BM n.v.; isotypes K!, SING!).

Terrestrial. Lower montane forest. Elevation: 1200–1700 m. Not endemic to Mount Kinabalu.

Additional material examined: EASTERN SHOULDER: 1200 m, *RSNB 300* (K, SING); LIWAGU RIVER TRAIL: 1500 m, *Parris & Croxall 9090* (K); LUBANG: *Topping 1794* (SING); MESILAU: 1200 m, *Clemens 51504* (K); PENIBUKAN: 1200 m, *Clemens 31029* (K); PINOSUK PLATEAU: 1500 m, *Parris & Croxall 9146* (K); TENOMPOK: 1500 m, *Clemens 28094* (K), 1400 m, *Holttum SFN 25394* (K, SING); WEST MESILAU RIVER: 1600–1700 m, *Beaman 7460* (MICH).

13.13.2. Polystichum holttumii C. Chr. in C. Chr. & Holttum, Gardens' Bull. 7: 256, pl. 53 (1934). Type: Paka-paka Cave: 3000 m, *Holttum SFN 25513* (holotype BM n.v.; isotypes K!, SING!).

Terrestrial. Lower and upper montane forest. Elevation: 2000–3300 m. Not endemic to Mount Kinabalu.

Additional material examined: GURULAU SPUR: 3200 m, *Clemens 50809* (K); KEMBURONGOH: 2100 m, *Clemens 27948* (K), 2900 m, *33222* (K), 2000–2100 m, *Meijer SAN 29122* (K); KEMBURONGOH/PAKA-PAKA CAVE: 2600 m, *Sinclair et al. 9106* (K, SING); LAYANG-LAYANG/PAKA-PAKA CAVE: 3000 m, *Edwards 2194* (K); MESILAU CAVE: 2000 m, *RSNB 2630* (K); PAKA-PAKA CAVE: 3000 m, *Clemens 27973* (K), 3100 m, *Edwards 2210* (K), 3100 m, *Parris 11577* (K), 3000 m, *Sinclair et al. 9171* (SING), *Topping 1704* (SING); PAKA-PAKA CAVE/LUMU-LUMU: 2400 m, *Clemens 28971* (K); PANAR LABAN: 3300 m, *Parris 11588* (K); SUMMIT TRAIL: 2600 m, *Parris 11563* (K), 3000 m, *Parris & Croxall 8704* (K), 3300 m, *8741* (K).

13.13.3. Polystichum kinabaluense C. Chr. in C. Chr. & Holttum, Gardens' Bull. 7: 255, pl. 52 (1934). Type: Summit Trail: *Clemens 10648* (lectotype of Parris, here designated, K!).

Polystichum clemensiae Copel., Philipp. J. Sci. 56: 474, t. 4 (1935). Type: Gurulau Spur: 3700 m, *Clemens 50864* (holotype PNH?†; isotype K!).

Terrestrial. Upper montane forest. Elevation: 3500–3800 m. Not endemic to Mount Kinabalu.

Additional material examined: MOUNT KINABALU: 3700 m, *Holttum SFN 25490* (K, SING); SUMMIT TRAIL: 3500 m, *Parris & Croxall 8772* (K), 3500 m, *8773* (K), *Topping 1702* (SING); UPPER KINABALU: *Clemens 27968* (K, SING); VICTORIA PEAK: 3800 m, *Clemens 51391* (K), 3800 m, *51542* (K).

13.13.4. Polystichum obtusum J. Sm. ex C. Presl, Epim., 53 (1849). C. Chr. & Holttum, Gardens' Bull. 7: 255 (1934).

Polystichum obtusum J. Sm. ex C. Presl var. *densum* Alderw., Bull. Jard. Bot. Buit. 2, 1: 12 (1911). C. Chr. & Holttum, Gardens' Bull. 7: 255 (1934).

Terrestrial. Lower montane forest. Elevation: 1700–2400 m.

Material examined: KEMBURONGOH: 2100 m, *Holttum SFN 25525b* (SING), 2100 m, *25534* (K, SING), 1800 m, *25714* (SING); LUMU-LUMU: 2100 m, *Clemens 29937* (K); MESILAU BASIN: 2100–2400 m, *Clemens 29629* (K); PIG HILL: 2000–2300 m, *Beaman 9836* (K, MICH, MSC); ULAR HILL TRAIL: 1700 m, *Parris 11457* (K), 1800 m, *Parris & Croxall 9089* (K).

13.13.5. Polystichum oppositum Copel., Philipp. J. Sci. 56: 474, t. 5 (1935). Type: Mesilau River: *Clemens s.n.* (holotype PNH?†), 2700 m, *51630* (isotype?, label details except number agree with protologue, K!).

Terrestrial. Montane forest. Elevation: 2700–2800 m. Not endemic to Mount Kinabalu.

Additional material examined: DACHANG: 2700 m, *Clemens 29053* (K); EASTERN SHOULDER: 2700 m, *RSNB 949b* (K, SING); GOKING'S VALLEY: 2800 m, *Fuchs 21481* (K); KINATEKI RIVER HEAD: 2700 m, *Clemens 31922* (K).

13.14. PTERIDRYS

Christensen, C. & Ching, R. C. 1934. *Pteridrys*, a new fern genus from tropical Asia. Bull. Fan Mem. Inst. Biol. Bot. 5: 125–145.

13.14.1. Pteridrys microthecia (Fée) C. Chr. & Ching, Bull. Fan Mem. Inst. Biol. Bot. 5: 139, t. 14, 18 (1934).

Dryopteris microthecia (Fée) C. Chr., Index Fil. Suppl. 2: 15 (1917). C. Chr. & Holttum, Gardens' Bull. 7: 243 (1934).

Terrestrial. Hill forest. Elevation: 900–1100 m.

Material examined: DALLAS: 900 m, *Clemens 27384* (US), 1100 m, *Holttum SFN 25261* (SING); KINATEKI RIVER (NEAR MINITINDUK): 900 m, *Holttum SFN 25584* (K); MINITINDUK: 900 m, *Holttum SFN 25584* (SING).

13.15. STENOLEPIA

Copeland, E. B. 1947. Genera Filicum 106–107.

13.15.1. Stenolepia tristis (Blume) Alderw., Bull. Dept. Agric. Ind. Neerl. 27: 45 (1909). C. Chr. & Holttum, Gardens' Bull. 7: 226 (1934).

Athyrium atropurpureum Copel., Philipp. J. Sci. 12C: 59 (1917). C. Chr. & Holttum, Gardens' Bull. 7: 267 (1934). Type: Low's Peak: *Clemens 10620* (lectotype of Copeland, in sched., MICH!).

Terrestrial. Upper montane forest and scrub. Elevation: 2100–3800 m.

Additional material examined: GURULAU SPUR: 3800 m, *Clemens 51392* (K); KEMBURONGOH/LUMU-LUMU: 2100 m, *Clemens 27946* (SING); MOUNT KINABALU: 3500 m, *Molesworth-Allen 3268* (K); PANAR LABAN: 3300 m, *Parris 11595* (K); SHANGRI LA VALLEY: 3400 m, *Collenette 21513* (L); SUMMIT TRAIL: 3300 m, *Parris & Croxall 8778* (K); UPPER KINABALU: 2100 m, *Clemens 27946* (K).

13.16. TECTARIA

Holttum, R. E. 1981. The fern genus *Tectaria* Cav. in Malaya. Gardens' Bull. 34: 132–147. Holttum, R. E. 1988. Studies in the fern genera allied to *Tectaria* Cav. VII. Species of *Tectaria* sect. *Sagenia* (Presl) Holttum in Asia excluding Malesia. Kew Bull. 43: 475–489. Parris, B. S., Jermy, A. C., Camus, J. M. & Paul, A. M. 1984. The Pteridophyta of Gunung Mulu National Park. In Studies on the Flora of Gunung Mulu National Park. Ed. A. C. Jermy: 219–220. Forest Dept., Kuching, Sarawak. Holttum, R. E. *Tectaria*. Fl. Males. 2, 2 (in press).

13.16.1. Tectaria angulata (Willd.) C. Chr., Index Fil. Suppl. 3: 177 (1934).

Tectaria trifolia (Alderw.) C. Chr. in C. Chr. & Holttum, Gardens' Bull. 7: 261 (1934).

Terrestrial. Elevation: 200–800 m.

Material examined: EASTERN SHOULDER: 800 m, *RSNB 580* (K, SING); KEBAYAU: 200 m, *Holttum s.n.* (SING).

13.16.2. Tectaria beccariana (Ces.) C. Chr., Index Fil. Suppl. 3: 177 (1934).

Tectaria vasta sensu C. Chr. & Holttum non (Blume) Copel., Gardens' Bull. 7: 261 (1934).

Terrestrial. Lowland dipterocarp forest. Elevation: 300–600 m.

Material examined: KEBAYAU/KAUNG: 300 m, *Holttum SFN 25107* (K, SING), *Topping 1910* (US); KIAU: *Topping 1540* (US); PORING HOT SPRINGS: 600 m, *Parris & Croxall 8998* (K).

13.16.3. Tectaria christii Copel., Philipp. J. Sci. 2C: 416 (1907).

Tectaria coadunata C. Chr. & Holttum non (Wall.) C. Chr., Gardens' Bull. 7: 260 (1934).

Pendulous on rocks. Lower montane forest along the Kinateki River above Minitinduk. Elevation: 1000 m. Based on *Holttum SFN 25587* (BM n.v., SING n.v.).

13.16.4. Tectaria crenata Cav., Descr., 250 (1802). C. Chr. & Holttum, Gardens' Bull. 7: 260 (1934).

Terrestrial. Lower montane forest. Elevation: 800–1800 m.

Material examined: EASTERN SHOULDER, CAMP 1: 800 m, *RSNB 1186* (K, SING); TENOMPOK: 1500 m, *Clemens 26931* (K), 1500 m, *29411* (SING), 1400 m, *Holttum SFN 25404* (SING); ULAR HILL TRAIL: 1700 m, *Parris 11437* (K), 1800 m, *Parris & Croxall 9086* (K).

13.16.5. Tectaria decurrens (C. Presl) Copel., Elmers Leaflets 1: 234 (1907). C. Chr. & Holttum, Gardens' Bull. 7: 260 (1934).

Terrestrial. Hill dipterocarp and lower montane forest. Elevation: 700–1500 m.

Material examined: KIAU: *Topping 1540* (SING); MELANGKAP KAPA: 700–1000 m, *Beaman 8801* (MICH, MSC); PORING HOT SPRINGS/LANGANAN WATER FALLS: 1000 m, *Parris & Croxall 8937* (K); TAHUBANG RIVER: 1200 m, *Clemens 31038* (K); TENOMPOK: 1500 m, *Clemens 26859* (K, SING).

13.16.6. Tectaria dissecta (G. Forster) Lellinger, Amer. Fern J. 58: 156 (1968).

Dryopteris dissecta (G. Forster) Kuntze, Rev. Gen. Pl. 2: 812 (1891). C. Chr. & Holttum, Gardens' Bull. 7: 253 (1934).

Terrestrial. Hill forest. Elevation: 900 m.

Material examined: KADAMAIAN RIVER NEAR MINITINDUK: 900 m, *Holttum SFN 25571* (K); MINITINDUK: 900 m, *Holttum SFN 25571* (SING).

13.16.7. Tectaria griffithii (Baker) C. Chr., Index Fil. Suppl. 3: 180 (1934).

Tectaria malayensis (Christ) Copel., Philipp. J. Sci. 2C: 416 (1907). C. Chr. & Holttum, Gardens' Bull. 7: 260 (1934).

Terrestrial. Hill forest and lower montane forest. Elevation: 300–1500 m.

Material examined: DALLAS: 900 m, *Clemens 26845* (K); KEBAYAU/KAUNG: 300 m, *Holttum SFN 25108* (SING); PENIBUKAN: 1200 m, *Clemens 32065* (K); TAHUBANG RIVER: 900 m, *Holttum SFN 25598* (K, SING); TENOMPOK: 1500 m, *Clemens 27329* (K).

13.16.8. Tectaria holttumii C. Chr. in C. Chr. & Holttum, Gardens' Bull. 7: 259, pl. 55 (1934). Type: Minitinduk: 900 m, *Holttum SFN 25570* (holotype BM n.v.; isotypes K!, SING!).

Terrestrial. Elevation: 900–1200 m. Not endemic to Mount Kinabalu.

Additional material examined: DALLAS: 900 m, *Clemens 27313* (SING); TENOMPOK: 900–1200 m, *Clemens 29516* (K).

13.16.9. Tectaria nitens Copel., Philipp. J. Sci. 56: 475, t. 6 (1935). Type: Penibukan: 1200 m, *Clemens 40652* (holotype PNH?†; isotype K!).

Terrestrial. Lower montane forest. Elevation: 1200 m. Endemic to Mount Kinabalu.

Ferns of Kinabalu

13.16.10. Tectaria pleiosora (Alderw.) C. Chr. in C. Chr. & Holttum, Gardens' Bull. 7: 260 (1934).

Terrestrial. Lowland forest. Elevation: 500–900 m.

Material examined: DALLAS: 900 m, *Clemens 27295* (SING), 800 m, *Holttum SFN 25726* (K, SING); KIPUNGIT FALLS: 600 m, *Parris & Croxall 8917* (K); PENATARAN RIVER: 500 m, *Beaman 9286* (MICH).

13.16.11. Tectaria trichotoma (Fée) Tagawa, Acta Phytotax. Geobot. 25: 180 (1973).

Tectaria balabacensis (Christ) M. G. Price, Contr. Univ. Mich. Herb. 16: 199 (1987).

Terrestrial. Hill forest on steep slopes on ultramafics. Elevation: 800–1000 m.

Material examined: LOHAN RIVER: 800–1000 m, *Beaman 9059* (K, MICH, MSC).

14. GLEICHENIACEAE

14.1. DICRANOPTERIS

14.1.1. Dicranopteris clemensiae Holttum, Reinwardtia 4: 275 (1957). Type: Mount Kinabalu: *Clemens 28745* (holotype BM n.v.; isotype K n.v.).

Terrestrial. Lower montane forest? Endemic to Mount Kinabalu; known only from the type.

14.1.2. Dicranopteris curranii Copel., Philipp. J. Sci. 81: 4 (1952).

Terrestrial. Lower montane forest margin. Elevation: 1600 m.

Material examined: LIWAGU RIVER TRAIL: 1600 m, *Parris & Croxall 9039* (K).

14.1.3. Dicranopteris linearis (N. L. Burm.) Underw.

a. var. **demota** Holttum, Reinwardtia 4: 275 (1957). Type: Tenompok: 1500 m, *Clemens 29535* (holotype K!; isotype BM n.v.).

Gleichenia linearis (N. L. Burm.) Clarke, Trans. Linn. Soc. Bot. 1: 428 (1880). C. Chr. & Holttum, Gardens' Bull. 7: 212 (1934) p.p.

Terrestrial. Lower montane forest. Elevation: 900–2200 m. Not endemic to Mount Kinabalu.

Additional material examined: DALLAS: 900 m, *Holttum s.n.* (SING); KIAU VIEW TRAIL: 1600 m, *Parris 11615* (K); SUMMIT TRAIL: 2200 m, *Parris & Croxall 8856* (K); TENOMPOK/LUMU-LUMU: 1500 m, *Holttum SFN 25438* (K, SING).

b. var. **montana** Holttum, Reinwardtia 4: 276 (1957).

Gleichenia linearis (N. L. Burm.) Clarke, Trans. Linn. Soc. Bot. 1: 428 (1880). C. Chr. & Holttum, Gardens' Bull. 7: 212 (1934) p.p.

Terrestrial. Lower montane forest, in clearings. Elevation: 1500–1700 m.

Material examined: KIAU VIEW TRAIL: 1600 m, *Parris 11620* (K), 1700 m, *Parris & Croxall 8552* (K); MARAI PARAI SPUR: 1500 m, *Holttum SFN 25617* (K, SING).

c. var. subpectinata (Christ) Holttum, Reinwardtia 4: 277 (1957).

Terrestrial. Hill forest. Elevation: 400 m.

Material examined: KAUNG: 400 m, *Holttum s.n.* (SING).

d. var. subspeciosa Holttum, Reinwardtia 4: 278 (1957). Type: Kiau: *Topping 1516* (holotype US n.v.; isotype SING!).

Terrestrial. Margin of lowland dipterocarp forest. Elevation: 500–1600 m. Not endemic to Mount Kinabalu.

Additional material examined: KIAU VIEW TRAIL: 1600 m, *Parris 11616* (K); PORING HOT SPRINGS: 500 m, *Parris & Croxall 8980* (K).

14.2. DIPLOPTERYGIUM

14.2.1. Diplopterygium brevipinnulum (Holttum) Parris, comb. nov.

Gleichenia brevipinnula Holttum, Reinwardtia 4: 264 (1957).

Terrestrial. Lower and upper montane ridge forest. Elevation: 2200–2900 m.

Material examined: EASTERN SHOULDER: 2900 m, *RSNB 926* (K, SING); KINATEKI RIVER HEAD: 2400 m, *Clemens 31994* (K); MESILAU TRAIL: 2400 m, *Meijer SAN 38590* (K); SUMMIT TRAIL: 2500 m, *Parris 11605* (K), 2600 m, *Parris & Croxall 8801* (K), 2200 m, *8859* (K).

14.2.2. Diplopterygium bullatum (T. Moore) Parris, comb. nov.

Gleichenia bullata T. Moore, Index, 374 (1862). C. Chr. & Holttum, Gardens' Bull. 7: 212 (1934). Type: Mount Kinabalu: *Low s.n.* (holotype K!; isotype CGE!).

Terrestrial. Lower and upper montane ridge forest. Elevation: 1800–2900 m. Not endemic to Mount Kinabalu.

Additional material examined: EASTERN SHOULDER: 2900 m, *RSNB 905* (K, SING); JANET'S HALT: 2400 m, *Collenette 562* (K); JANET'S HALT/SHEILA'S PLATEAU: 2900 m, *Collenette 21557* (K); KEMBURONGOH: 2100 m, *Mikil SAN 29225* (K), 2400 m, *Sinclair et al. 9095* (K); KEMBURONGOH/LUMU-LUMU: 1800 m, *Holttum SFN 25450* (K, SING); KEMBURONGOH/ PAKA-PAKA CAVE: 2400 m, *Sinclair et al. 9095* (SING); LAYANG-LAYANG: 2600 m, *Edwards 2200* (K); LETENG RIVER (NEAR KUNDASANG): 1800 m, *Meijer SAN 24122* (K); LUBANG/PAKA-PAKA CAVE: *Clemens 10724* (K); MENTEKI RIDGE: 2400 m, *RSNB 7143* (K); PAKA-PAKA CAVE: 2100–2400 m, *Meijer SAN 22072* (K); SUMMIT TRAIL: 2500–2900 m, *Jacobs 5728* (K), 2600 m, *Parris 11553* (K), 2500 m, *Parris & Croxall 8810* (K).

14.2.3. Diplopterygium longissimum (Blume) Nakai, Bull. Nat. Sci. Mus. Tokyo 29: 53 (1950).

Gleichenia glauca (Thunb.) Hook. var. *longissima* (Blume) ined., (no basionym was cited) C. Chr. & Holttum, Gardens' Bull. 7: 212 (1934).

Terrestrial. Lower montane forest. Elevation: 1600–1800 m.

Material examined: KEMBURONGOH/LUMU-LUMU: 1800 m, *Holttum SFN 25448* (K, SING); PARK HEADQUARTERS: 1600 m, *Parris & Croxall 9124* (K).

14.2.4. Diplopterygium norrisii (Mett.) Nakai, Bull. Nat. Sci. Mus. Tokyo 29: 54 (1950).

Gleichenia norrisii Mett. in Kuhn, Linnaea 36: 165 (1869). C. Chr. & Holttum, Gardens' Bull. 7: 212 (1934). *Gleichenia norrisii* Mett. in Kuhn var. *floccigera* C. Chr. in C. Chr. & Holttum, Gardens' Bull. 7: 212 (1934). Type: Gurulau Spur: *Gibbs 3976* (syntype BM n.v.); Kiau: *Clemens 9986* (syntype BM? n.v.; isosyntype K!).

Terrestrial. Lower montane forest. Elevation: 1400–1600 m.

Additional material examined: GURULAU SPUR: *Topping 1633* (SING); KIAU VIEW TRAIL: 1600 m, *Parris & Croxall 9127* (K); TENOMPOK: 1500 m, *Clemens 28267* (K), 1500 m, *29430* (K), 1400 m, *Holttum SFN 25405* (K, SING).

14.3. GLEICHENIA

14.3.1. Gleichenia dicarpa R. Br., Prodr., 161 (1810).

Terrestrial. Lower montane ridge forest. Elevation: 2400–2600 m.

Material examined: SUMMIT TRAIL: 2400 m, *Parris 11543* (K), 2600 m, *Parris & Croxall 8699* (K), 2500 m, *8818* (K).

14.3.2. Gleichenia microphylla R. Br., Prodr., 161 (1810).

Gleichenia microphylla R. Br. var. *semivestita* (Labill.) Alderw., Handb. Malayan Ferns Suppl., 80 (1917). C. Chr. & Holttum, Gardens' Bull. 7: 210 (1934) p.p.

Terrestrial. Lower montane forest, probably on ultramafics. Elevation: 1800 m.

Material examined: MARAI PARAI: *Topping 1870* (SING); MARAI PARAI SPUR: *Clemens 10947* (K); PENATARAN BASIN: 1800 m, *Clemens 40189* (K).

14.3.3. Gleichenia peltophora Copel., Philipp. J. Sci. 40: 292, t. 1 (1929).

a. var. peltophora

Gleichenia circinnata Sw. var. *borneensis* Baker, J. Bot. 17: 37 (1879). Type: Mount Kinabalu: 1500–1800 m, *Burbidge s.n.* (holotype K!). *Gleichenia borneensis* (Baker) C. Chr. in C. Chr. & Holttum, Gardens' Bull. 7: 211 (1934).

Terrestrial. Lower and upper montane forest, in open areas. Elevation: 1500–3000 m.

Additional material examined: EASTERN SHOULDER: 2700 m, *RSNB 995* (K, SING); KEMBURONGOH: 2200 m, *Sinclair et al. 9067* (K); KEMBURONGOH/PAKA-PAKA CAVE: 2700 m, *Holttum SFN 25517* (K, SING), 2200 m, *Sinclair et al. 9067* (SING); LAYANG-LAYANG: 2600 m, *Parris & Croxall 8698* (K); MESILAU HILL: 2100 m, *Poore H 408* (K); PAKA-PAKA CAVE: 2700–3000 m, *Clemens 27470* (K).

14.3.4. Gleichenia vulcanica Blume, Enum. Pl. Javae, 251 (1828).

Gleichenia microphylla R. Br., Gardens' Bull. 7: 210 (1934).

Terrestrial. Lower and upper montane forest, in open areas. Elevation: 2000–3200 m.

Material examined: KEMBURONGOH: *Clemens 10533* (K), 2700 m, *28015* (K); KEMBURONGOH/PAKA-PAKA CAVE: 2600 m, *Sinclair et al. 9202* (K, SING); MESILAU CAVE: 2000–2100 m, *Beaman 8145* (MICH); MOUNT KINABALU: 3200 m, *Haviland 1483* (K); MT. TIBABAR: 2100 m, *Sinclair et al. 9039* (K, SING); PAKA-PAKA CAVE: 3000 m, *Holttum SFN 25510* (K, SING), 2400 m, *Meijer SAN 22071* (K); SUMMIT TRAIL: 2500–2900 m, *Jacobs 5725* (K).

14.4. STICHERUS

Parris, B. S. 1982. A new species of *Gleichenia* from lowland forest. In Notulae et Novitates Muluensis No. 2, ed. A. C. Jermy & K. P. Kavanagh. Bot. J. Linn. Soc. 85: 30–31.

14.4.1. Sticherus hirtus (Blume) Ching

a. var. paleaceus (Baker) Parris, comb. nov.

Gleichenia vestita Blume var. *paleacea* Baker, J. Bot. 17: 38 (1879). *Gleichenia hirta* Blume var. *paleacea* (Baker) C. Chr. in C. Chr. & Holttum, Gardens' Bull. 7: 212 (1934).

Terrestrial. Lower and upper montane forest. Elevation: 1200–2900 m.

Material examined: GOKING'S VALLEY: 2800 m, *Fuchs 21490* (K); JANET'S HALT/SHEILA'S PLATEAU: 2900 m, *Collenette 21558* (K); KEMBURONGOH: 2700 m, *Clemens 28016* (K), 2100 m, *Sinclair et al. 9054* (K, SING); KEMBURONGOH/LUMU-LUMU: 1800 m, *Holttum SFN 25449* (K, SING); KINATEKI RIVER: 1200–1500 m, *Clemens 31392* (K); LETENG RIVER (NEAR KUNDASANG): 1800 m, *Meijer SAN 24122* (K); LUBANG/PAKA-PAKA CAVE: *Topping 1753* (SING); MESILAU TRAIL: 2400 m, *Meijer SAN 38595* (K); PARK HEADQUARTERS: *Edwards 2237* (K); SUMMIT TRAIL: 2200 m, *Parris & Croxall 8857* (K); TENOMPOK: 1500 m, *Clemens 28785* (K).

14.4.2. Sticherus impressus (Parris) Parris, comb. nov.

Gleichenia impressa Parris in Jermy & Kavan., Bot. J. Linn. Soc. 85: 30 (1982).

Terrestrial. Hill dipterocarp forest. Elevation: 1000 m.

Material examined: PORING HOT SPRINGS/LANGANAN WATER FALLS: 1000 m, *Parris & Croxall 8924* (K).

14.4.3. Sticherus loheri (Christ) Copel.

a. var. major (Holttum) Parris, comb. nov.

Gleichenia loheri Christ var. *major* Holttum, Reinwardtia 4: 272 (1957).

Terrestrial. Upper montane forest. Elevation: 2400–2800 m.

Material examined: LUBANG/PAKA-PAKA CAVE: *Topping 1755* (SING); SUMMIT TRAIL: 2400 m, *Parris 11511* (K), 2800 m, *Parris & Croxall 8834* (K), 2700 m, *8837* (K).

14.4.4. Sticherus truncatus (Willd.) Nakai, Bull. Nat. Sci. Mus. Tokyo 29: 20 (1950).

a. var. truncatus

Gleichenia laevigata (Willd.) Hook., Sp. Fil. 1: 10 (1844).

Terrestrial. Lowland dipterocarp forest to lower montane forest. Elevation: 300–2200 m.

Material examined: KAUNG: 300 m, *Holttum s.n.* (SING); KIAU: *Gibbs 3966* (K); KIPUNGIT FALLS TRAIL: 600 m, *Parris & Croxall 8910* (K); SUMMIT TRAIL: 2200 m, *Parris & Croxall 8858* (K).

15. GRAMMITIDACEAE

15.1. ACROSORUS

Parris, B. S. 1986. Grammitidaceae of Peninsular Malaysia and Singapore. Kew Bull. 41: 507–508.

15.1.1. Acrosorus friderici-et-pauli (Christ) Copel., Philipp. J. Sci. 1 Suppl. 2: 159 (1906).

Polypodium streptophyllum sensu C. Chr. & Holttum non Baker, Gardens' Bull.
7: 298 (1934).

Epiphyte. Lower montane ridge forest. Elevation: 1500–2700 m.

Material examined: GURULAU SPUR: 1700 m, *Gibbs 4017* (BM); KEMBURONGOH: 2100–2700
m, *Clemens 27992* (K); KEMBURONGOH/LUMU-LUMU: 1800 m, *Holttum SFN 25473* (BM, K, SING);
KILEMBUN BASIN: 1700 m, *Clemens 33677* (K, SING), 1500 m, *40013* (K, SING); KILEMBUN RIVER:
1800 m, *Clemens 32473* (BO); LUMU-LUMU: 1800–2100 m, *Enriquez SFN 18163* (SING); MURU-TURA
RIDGE: 1500 m, *Clemens 34222* (K, SING), 1500–1800 m, *34348* (K, SING); PENIBUKAN: 1500 m,
Clemens 30985 (K); SUMMIT TRAIL: 1900 m, *Parris & Croxall 8497* (K); TENOMPOK/TOMIS:
Clemens 29506 (BM,K).

15.1.2. Acrosorus streptophyllus (Baker) Copel., Philipp. J. Sci. 56: 480 (1935).

Polypodium streptophyllum Baker, J. Bot. 17: 42 (1879).

Epiphyte. Hill dipterocarp ridge forest and lower montane oak ridge forest.
Elevation: 1000–1900 m.

Material examined: PORING HOT SPRINGS/LANGANAN WATER FALLS: 1000 m, *Parris &
Croxall 8958* (K); SUMMIT TRAIL: 1900 m, *Parris & Croxall 8498* (K).

15.2. CALYMMODON

Copeland, E. B. 1927. The genus *Calymmodon*. Philipp. J. Sci. 34: 259–271.

15.2.1. Calymmodon clavifer (Hook.) T. Moore, Index, 219 (1861).

Grammitis clavifer Hook., Second Cent. Ferns, t. 5 (1860). Type: Mount
Kinabalu: *Low s.n.* (lectotype of Parris, here designated, K!; isolectotype CGE!).
Polypodium clavifer (Hook.) Hook., Species Filicum 4: 176 (1862). C. Chr. &
Holttum, Gardens' Bull. 7: 298 (1934).

Epiphyte. Lower montane forest. Elevation: 1200–2700 m. Not endemic to
Mount Kinabalu.

Additional material examined: KEMBURONGOH: *Clemens 28000* (K), 2400 m, *28315* (BM, K),
2100 m, *Gibbs 4140* (BM, K); KEMBURONGOH/LUMU-LUMU: 2100–2700 m, *Clemens 28979* (BM,
K), 1800 m, *Holttum SFN 25465* (BM, K, SING), 1800 m, *Sinclair et al. 9028* (K, SING);
KEMBURONGOH/PAKA-PAKA CAVE: 2700 m, *Clemens 29104A* (BM); KINATEKI RIVER: 1200–
1500 m, *Clemens 31423* (K); LIWAGU RIVER TRAIL: 1800 m, *Parris & Croxall 9070* (K); LUMU-
LUMU: 1800 m, *Clemens 27065A* (BM); MARAI PARAI: 1800–2400 m, *Clemens 32958* (K, SING);
MARAI PARAI SPUR: *Clemens 11043* (BM, K); SUMMIT TRAIL: 1800 m, *Parris & Croxall 8496* (K),
1800 m, *8871* (K).

15.2.2. Calymmodon cucullatus (Nees & Blume) C. Presl, Tent., 204 (1836).

Polypodium cucullatum Nees & Blume, Nova Acta 11: 121, pl. 12, f. 3 (1823).
Polypodium cucullatum Nees & Blume var. *subgracillimum* (Alderw.) C. Chr. in C.
Chr. & Holttum p.p., Gardens' Bull. 7: 297 (1934).

Epiphyte. Lower and upper montane forest. Elevation: 1500–3000 m.

Material examined: DACHANG: *Clemens 29058* (BM, K); GURULAU SPUR: 2400 m, *Clemens
51191* (K), 1700 m, *Gibbs 4018* (K); KEMBURONGOH/PAKA-PAKA CAVE: *Topping 1659* (SING, US);
LUBANG/PAKA-PAKA CAVE: *Topping 1726* (US); MARAI PARAI: 1500 m, *Clemens 33161* (SING);
MESILAU TRAIL: 1600 m, *Hale 29040* (US); MOUNT KINABALU: 2700 m, *E. F. Allen s.n.* (K);
PAKA-PAKA CAVE: 3000 m, *Holttum SFN 25500* (K, SING), *Topping 1666* (US), *1713* (US); SUMMIT
TRAIL: 2400 m, *Parris & Croxall 8488* (K), 2400 m, *8641* (K), 2500 m, *8824* (K), 2800 m, *8830* (K), 2400
m, *8895* (K), 2300 m, *9167* (K).

15.2.3. Calymmodon gracilis (Fée) Copel., Philipp. J. Sci. 34: 266 (1927).

Polypodium consociatum Alderw., Bull. Jard. Bot. Buit. 2, 7: 41, pl. 4 (1912). C. Chr. & Holttum, Gardens' Bull. 7: 298 (1934).

Epiphyte. Lower montane to lower levels of upper montane forest. Elevation: 1200–3000 m.

Material examined: EASTERN SHOULDER: 3000 m, *RSNB 722* (K, SING); EASTERN SHOULDER, CAMP 4: 2700 m, *RSNB 1145* (K); GURULAU SPUR: *Topping 1816* (BM); KEMBURONGOH: 2100 m, *Price 239* (K); LIWAGU RIVER TRAIL: 1400 m., Edwards *2176A* (K), *2177* (K); LUGAS HILL: 1300 m, *Beaman 10546* (MICH); MARAI PARAI SPUR: 1300 m, *Parris 11525* (K); MEMPENING TRAIL: 1600 m, *Parris 11514* (K); MOUNT KINABALU: 1500 m, *Cox 2550a* (K), 1500 m, *Shim SAN 75447* (SAN); MT. LENAU (BETWEEN TENOMPOK AND KEMBURONGOH): 1600 m, *Sinclair et al. 9008* (K, SING); PARK HEADQUARTERS: 1500 m, *Parris & Croxall 8478* (K); PENIBUKAN: 1200–1500 m, *Clemens 31423A* (K), 1200 m, *51709* (K); SUMMIT AREA: *Clemens 30461* (K); SUMMIT TRAIL: 2600 m, *Parris 11552* (K), 2500 m, *Parris & Croxall 8635* (K), 1800 m, *8651* (K), 2700 m, *8719* (K), 2500 m, *8797* (K), 1800 m, *8870* (K), 1900 m, *8894* (K); TENOMPOK/LUMU-LUMU: 1500 m, *Holttum SFN 25428* (BM, K, SING).

15.2.4. Calymmodon hygroscopicus Copel., Philipp. J. Sci. 34: 265, t. 5 (1927).

Epiphyte. Lower montane forest. Elevation: 1500–2400 m.

Material examined: LIWAGU RIVER TRAIL: 1500 m, *Parris & Croxall 8484* (K); SUMMIT TRAIL: 2000 m, *Parris & Croxall 8494* (K), 2100 m, *8650* (K), 2000 m, *8892* (K), 2100 m, *8893* (K), 2200 m, *8896* (K), 2300 m, *8897* (K), 2400 m, *8898* (K).

15.2.5. Calymmodon muscoides (Copel.) Copel., Philipp. J. Sci. 34: 264 (1927).

Epiphyte. Lower montane ridge and valley forest. Elevation: 1500 m.

Material examined: LIWAGU RIVER TRAIL: 1500 m, *Parris & Croxall 9041* (K); MEMPENING TRAIL: 1500 m, *Parris & Croxall 9116* (K).

15.2.6. Calymmodon sp. 1

Epiphyte. Lower and upper montane forest. Elevation: 1500–3100 m. Not endemic to Mount Kinabalu.

Material examined: PAKA-PAKA CAVE: 3100 m, *Parris 11663* (K), 3000 m, *Parris & Croxall 8700* (K), 3000 m, *8784* (K); PARK HEADQUARTERS: 1500 m, *Parris 11432* (K); SUMMIT TRAIL: 2800 m, *Parris & Croxall 8518* (K), 2300 m, *8637* (K), 2300 m, *8639* (K), 2400 m, *8640* (K), 2000 m, *8656* (K), 2500 m, *8798* (K), 2500 m, *8823* (K), 2800 m, *8831* (K), 2900 m, *8833* (K).

15.2.7. Calymmodon sp. 2

Epiphyte? Lower and upper montane forest. Elevation: 1500–2900 m. Not endemic to Mount Kinabalu.

Material examined: MARAI PARAI: 1500 m, *Clemens 39161* (K); PINOSUK PLATEAU: 1700 m, *RSNB 1869* (K); SUMMIT TRAIL: 2500–2900 m, *Jacobs 5714* (K).

15.2.8. Calymmodon sp. 3

Epiphyte. Lower montane forest. Elevation: 1500–2400 m. Not endemic to Mount Kinabalu.

Material examined: MARAI PARAI: *Clemens 11044* (BM); MEMPENING TRAIL: 1600 m, *Parris 11661* (K); PARK HEADQUARTERS: 1600 m, *Parris & Croxall 9123* (K), 1500 m, *9166* (K); SUMMIT TRAIL: 2400 m, *Jacobs 5774* (K), 1800–2000 m, *5792* (K).

15.2.9. Calymmodon sp. 4

Epiphyte. Lower montane oak and upper montane forest. Elevation: 1500–1800 m. Not endemic to Mount Kinabalu.

Material examined: KIAU VIEW TRAIL: 1600 m, *Parris & Croxall 8566* (K); KILEMBUN BASIN: 1500 m, *Clemens s.n.* (K); KILEMBUN RIVER HEAD: 1800 m, *Clemens 32467* (K, SING); SUMMIT TRAIL: 3000 m, *Edwards 2214* (K).

15.2.10. Calymmodon sp. 5

Polypodium cucullatum sensu C. Chr. & Holttum p.p. non Nees & Blume, Gardens' Bull. 7: 297 (1934). *Polypodium cucullatum* Nees & Blume var. *subgracillimum* (Alderw.) C. Chr. in C. Chr. & Holttum p.p., Gardens' Bull. 7: 297 (1934).

Epiphytic and terrestrial. Upper montane forest. Elevation: 1400–3800 m. Endemic to Mount Kinabalu.

Material examined: KEMBURONGOH: 2100 m, *Clemens 28085* (BM); KEMBURONGOH/LUMU-LUMU: 1800 m, *Holttum SFN 25732* (BM, K); MAMUT COPPER MINE: 1400 m, *Collenette 1022* (K); MESILAU BASIN: 3400 m, *Smith 532* (L); MOUNT KINABALU: 3200 m, *Haviland 1482* (K), 3500 m, *Molesworth-Allen 3265* (K); PAKA-PAKA CAVE: *Topping 1713* (BM); PANAR LABAN: 3300 m, *Parris 11590* (K); SUMMIT TRAIL: 3000 m, *Parris & Croxall 8506* (K), 3200 m, *8517* (K), 3200 m, *8738* (K), 2800 m, *8829* (K); VICTORIA PEAK: 3800 m, *Clemens 50865* (K).

15.2.11. Calymmodon sp. 6

Epiphyte. Lower and upper montane forest. Elevation: 1500–2800 m. Endemic to Mount Kinabalu.

Material examined: PARK HEADQUARTERS: 1500 m, *Parris & Croxall 8716* (K), 1500 m, *9022* (K); SUMMIT TRAIL: 2300 m, *Parris & Croxall 8636* (K), 2300 m, *8638* (K), 2500 m, *8826* (K), 2600 m, *8827* (K), 2600 m, *8828* (K), 2800 m, *8832* (K).

15.2.12. Calymmodon sp. 7

Epiphyte. Lower montane forest. Elevation: 2400 m. Endemic to Mount Kinabalu.

Material examined: SUMMIT TRAIL: 2400 m, *Parris & Croxall 8646* (K).

15.3. CTENOPTERIS

Copeland, E. B. 1953. Grammitidaceae of New Guinea. Philipp. J. Sci. 81: 99–115. Copeland, E. B. 1960. Fern Flora of the Philippines. Vol. 3: 525–535. Parris, B. S. 1986. Grammitidaceae of Peninsular Malaysia and Singapore. Kew Bull. 41: 492–499. Parris, B. S., Jermy, A. C., Camus, J. M., & Paul, A. M. 1984. The Pteridophyta of Gunung Mulu National Park. *In* Studies on the Flora of Gunung Mulu National Park, Sarawak. Ed. A. C. Jermy: 201. Forest Dept., Kuching, Sarawak.

15.3.1. Ctenopteris barathrophylla (Baker) Parris, Fern Gaz. 12: 118 (1980).

Epiphyte. Lower montane forest. Elevation: 1500 m.

Material examined: PENIBUKAN: 1500 m, *Clemens 51702* (K).

15.3.2. Ctenopteris aff. barathrophylla (Baker) Parris

Polypodium khasyanum sensu C. Chr. & Holttum non Hook., Gardens' Bull. 7: 303 (1934).

Epiphyte. Lower montane forest. Elevation: 2100 m. Not endemic to Mount Kinabalu.

Material examined: MESILAU BASIN: 2100 m, *Clemens 29029* (K,SING); PENIBUKAN: *Clemens 30748* (SING).

15.3.3. Ctenopteris blechnoides (Grev.) W. H. Wagner & Grether, Univ. Calif. Publ. Bot. 23: 61 (1948).

Polypodium moultonii Copel., Philipp. J. Sci. 10C: 149 (1915). C. Chr. & Holttum, Gardens' Bull. 7: 300 (1934).

Epiphyte. Hill dipterocarp forest. Elevation: 1000 m.

Material examined: PORING HOT SPRINGS/LANGANAN WATER FALLS: 1000 m, *Parris & Croxall 8521* (K).

15.3.4. Ctenopteris brevivenosa (Alderw.) Holttum, Rev. Fl. Mal. 2: 228, f. 126 (1955).

Polypodium mollicomum sensu Christensen & Holttum p.p. non Nees & Blume, Gardens' Bull. 7: 301 (1934).

Epiphyte. Lower montane forest. Elevation: 1400–3000 m.

Material examined: EASTERN SHOULDER, CAMP 4: 2700 m, *RSNB 1144* (K, SING); KEMBURONGOH: 2400–3000 m, *Clemens s.n.* (BM); KEMBURONGOH/LUMU-LUMU: 1800–2400 m, *Clemens 28968* (BM), 1800 m, *Holttum SFN 25461* (K, SING); KEMBURONGOH/PAKA-PAKA CAVE: 2700 m, *Molesworth-Allen 3258* (K); KUNDASANG: 1400 m, *Cox 2527* (K); LIWAGU RIVER TRAIL: 1400 m, *Edwards 2216* (K); MARAI PARAI: 1500 m, *Clemens 33212* (K); MARAI PARAI SPUR: 2700–3000 m, *Clemens 33212* (SING); PAKA-PAKA CAVE: 3000 m, *Gibbs 4251* (BM); PARK HEADQUARTERS: 1500 m, *Parris & Croxall 8476* (K), *Price 138* (K); SUMMIT TRAIL: 2400 m, *Edwards 2220* (K), 2500–2900 m, *Jacobs 5729* (K), *5795* (K), 2700 m, *Parris 11564* (K), 1900 m, *Parris & Croxall 8499* (K), 2600 m, *8661* (K), 2700 m, *8718* (K).

15.3.5. Ctenopteris celebica (Blume) Copel., Univ. Calif. Publ. Bot. 18: 225 (1942).

Polypodium celebicum Blume, Enum. Pl. Javae, 127 (1828). C. Chr. & Holttum, Gardens' Bull. 7: 303 (1934).

Epiphyte. Lower montane forest. Elevation: 1500–2100 m.

Material examined: KEMBURONGOH: 2100 m, *Holttum SFN 25524* (BM, K, SING); KILEMBUN RIVER HEAD: 1800 m, *Clemens 32467A* (SING); LIWAGU RIVER TRAIL: 1800 m, *Parris & Croxall 9062* (K); LUBANG: *Topping 1770* (SING, US); MARAI PARAI: 1500 m, *Clemens s.n.* (K), 1800 m, *32446* (K), 1800 m, *32869* (K, SING); MESILAU BASIN: 2100 m, *Clemens 29044* (K); MESILAU BASIN/LUMU LUMU: 2100 m, *Clemens 29912* (BM, K); SUMMIT TRAIL: 1800 m, *Parris & Croxall 8876* (K); ULAR HILL TRAIL: 1800 m, *Parris & Croxall 8528* (K).

15.3.6. Ctenopteris curtisii (Baker) Copel., Philipp. J. Sci. 81: 103 (1953).

Polypodium curtisii Baker, J. Bot. 19: 367 (1881). C. Chr. & Holttum, Gardens' Bull. 7: 304 (1934).

Terrestrial. Lower montane forest. Elevation: 1500–2400 m.

Material examined: KEMBURONGOH: 2400 m, *Clemens 28966* (BM, K); LIWAGU RIVER TRAIL: 1600 m, *Parris & Croxall 9061* (K); LUMU-LUMU: 1800 m, *Clemens 27965* (K); MEMPENING TRAIL: 1600 m, *Parris 11504* (K); MENTEKI RIVER: 1900 m, *Kitayama K665 UKMS 3828* (SAN); MOUNT KINABALU: *Shim SAN 75444* (SAN); PARK HEADQUARTERS: 1500 m, *Parris & Croxall 8472* (K); TENOMPOK: 1500 m, *Clemens 28966b* (K); TENOMPOK/LUMU-LUMU: 1500 m, *Holttum SFN 25432* (BM, K, SING).

15.3.7. Ctenopteris denticulata (Blume) C. Chr. & Tardieu-Blot, Not. Syst. (Paris) 8: 181 (1939).

Polypodium denticulatum (Blume) C. Presl, Tent., 178 (1836). C. Chr. & Holttum, Gardens' Bull. 7: 300 (1934).

Epiphyte. Lower montane forest. Elevation: 1200–1800 m.

Material examined: DALLAS/TENOMPOK: 1200 m, *Clemens 28157* (BM); GURULAU SPUR: *Clemens 10855* (BM); KIAU VIEW TRAIL: 1700 m, *Parris & Croxall 8523* (K); PINOSUK PLATEAU: 1600 m, *Parris & Croxall 9149* (K); POWER STATION: 1800 m, *Parris & Croxall 9069* (K); TENOMPOK: 1500 m, *Holttum SFN 25731* (BM, SING).

15.3.8. Ctenopteris fuscata (Blume) Kunze, Bot. Zeit. 4: 425 (1846).

Polypodium mollicomum sensu Christensen & Holttum p.p. non Nees & Blume, Gardens' Bull. 7: 301 (1934).

Epiphyte. Tall and short upper montane forest. Elevation: 3000–3800 m.

Material examined: EASTERN SHOULDER: 3000 m, *RSNB 723* (K, SING); MARAI PARAI: 3000 m, *Clemens 33205* (K); MARAI PARAI SPUR: 3000 m, *Clemens 33205* (SING); MESILAU BASIN: 3400 m, *Kitayama K743 UKMS 3904* (SAN); PAKA-PAKA CAVE: 3100 m, *Parris 11584* (K), 3000 m, *Parris & Croxall 8717* (K), *Topping 1714* (BM, SING); PANAR LABAN: 3300 m, *Parris 11670* (K); SUMMIT AREA: 3800 m, *Collenette 617* (K); SUMMIT TRAIL: 3100 m, *Edwards 2213* (K), 3500 m, *Molesworth-Allen 3282* (K), 3000 m, *Parris & Croxall 8504* (K), 3000 m, *8702* (K); UPPER KINABALU: *Clemens 27756* (K, SING).

15.3.9. Ctenopteris halconensis (Copel.) Copel., Fern Fl. Philip. 3: 533 (1960).

Polypodium halconense Copel., Philipp. J. Sci. 2C: 138, t. 2, f. 6 (1907). C. Chr. & Holttum, Gardens' Bull. 7: 304 (1934).

Epiphyte. Lower montane forest. Elevation: 2100 m.

Material examined: LUMU-LUMU: 2100 m, *Clemens 30447* (SING).

15.3.10. Ctenopteris millefolia (Blume) Copel., Gen. Fil., 219 (1947).

Polypodium millefolium Blume, Enum. Pl. Javae, 134 (1828). C. Chr. & Holttum, Gardens' Bull. 7: 304 (1934).

Terrestrial. Lower montane forest. Elevation: 1800 m.

Material examined: MARAI PARAI SPUR: *Clemens 11058* (BM); SUMMIT TRAIL: 1800 m, *Parris & Croxall 8877* (K).

15.3.11. Ctenopteris minuta (Blume) Holttum, Rev. Fl. Mal. 2: 228 (1955).

Epiphyte. Lower montane ridge forest. Elevation: 1400 m.

Material examined: LIWAGU RIVER TRAIL: 1400 m, *Parris & Croxall 8481* (K); MARAI PARAI SPUR: 1400 m, *Parris 11673* (K).

15.3.12. Ctenopteris mollicoma (Nees & Blume) Kunze, Bot. Zeit. 4: 425 (1846).

Polypodium mollicomum Nees & Blume, Nova Acta 11: 121, t. 12, f. 2 (1823). C. Chr. & Holttum, Gardens' Bull. 7: 301 (1934) p.p.

Epiphyte. Lower montane to upper montane forest. Elevation: 1500–3200 m.

Material examined: EASTERN SHOULDER, CAMP 4: 2900 m, *RSNB 1152* (K, SING); GURULAU SPUR: 3200 m, *Clemens 50901* (K), 2700 m, *51125* (K); KEMBURONGOH/LUMU-LUMU: 1800–2400 m, *Clemens 27994* (BM, K), 2100–2400 m, *28967* (BM), *28969* (K); MAMUT COPPER MINE: 1600–1700 m, *Beaman 9943* (MICH, MSC); MARAI PARAI SPUR: *Clemens 11049* (UC); PARK HEADQUARTERS: 1500 m, *Parris & Croxall 9023* (K); SUMMIT TRAIL: 2600 m, *Parris 11555* (K), 2100 m, *Parris & Croxall 8495* (K), 2500 m, *8649* (K), 2400 m, *9170* (K).

15.3.13. Ctenopteris multicaudata (Copel.) Copel., Philipp. J. Sci. 81: 112 (1953).

Epiphyte. Lower montane valley forest. Elevation: 1500–2100 m.

Material examined: MESILAU BASIN: 2100 m, *Clemens 29044* (BM); MESILAU RIVER: 1500 m, *RSNB 1362* (K, SING).

15.3.14. Ctenopteris nutans (Blume) J. Sm., Hist. Fil., 185 (1875).

Polypodium nutans Blume, Enum. Pl. Javae, 128 (1828). C. Chr. & Holttum, Gardens' Bull. 7: 302 (1934).

Epiphyte. Upper montane tall and short forest. Elevation: 2500–3400 m.

Material examined: DACHANG: 2900 m, *Clemens 28441* (K); MESILAU BASIN: 3400 m, *Smith 530* (L); PAKA-PAKA CAVE: *Clemens 28978* (BM, K, SING), 3000 m, *Parris & Croxall 8783* (K); PANAR LABAN: 3300 m, *Parris 11594* (K); SUMMIT TRAIL: 3100 m, *Edwards 2212* (K), 3000 m, *Parris & Croxall 8507* (K), 2500 m, *8644* (K), 3300 m, *8737* (K), 3300 m, *8740* (K), 3200 m, *8752* (K), 2500 m, *8796* (K).

15.3.15. Ctenopteris obliquata (Blume) Copel., Philipp. J. Sci. 81: 111 (1953).

Polypodium obliquatum Blume, Enum. Pl. Javae, 128 (1828). C. Chr. & Holttum, Gardens' Bull. 7: 303 (1934). *Prosaptia obliquata* (Blume) Mett., Novara Reise Bot. 1: 214 (1870).

Epiphyte. Lower montane valley forest. Elevation: 1400–1700 m.

Material examined: LIWAGU RIVER: 1500 m, *Parris & Croxall 8479* (K), 1400 m, *9029* (K); LIWAGU RIVER TRAIL: 1500 m, *Parris 11492* (K); WEST MESILAU RIVER: 1600–1700 m, *Beaman 8649* (K, MICH, MSC).

15.3.16. Ctenopteris papillata (Alderw.) Parris, comb. nov.

Polypodium papillatum Alderw., Bull. Jard. Bot. Buit. 2, 16: 35, t. 7 (1914). C. Chr. & Holttum, Gardens' Bull. 7: 299 (1934).

Epiphyte. Lower montane forest. Elevation: 1500 m.

Material examined: MARAI PARAI SPUR: 1500 m, *Holttum SFN 25604* (BM, K, SING).

15.3.17. Ctenopteris repandula (Mett.) C. Chr. & Tardieu-Blot in Lecompte, Fl. Gen. Indo-Ch. 7, 2: 533 (1941).

Epiphyte. Elevation: 800 m.

Material examined: HEMPUEN HILL: 800 m, *Abbe et al. 9968* (K).

15.3.18. Ctenopteris subminuta (Alderw.) Holttum, Rev. Fl. Mal. 2: 228, f. 127 (1955).

Polypodium subminutum Alderw., Handb. Malayan Ferns, 598 (1908). C. Chr. & Holttum, Gardens' Bull. 7: 301 (1934).

Epiphyte. Lower montane forest. Elevation: 1200–2000 m.

Material examined: KEMBURONGOH: 2000 m, *Molesworth-Allen s.n.* (K); KIAU VIEW TRAIL: 1600 m, *Parris 10810* (K); LUBANG: *Holttum SFN 25550* (BM, K, SING); MEMPENING TRAIL: 1600 m, *Parris 11515* (K); PARK HEADQUARTERS: 1500 m, *Parris & Croxall 8471* (K), 1500 m, *9026* (K); TENOMPOK/LUMU-LUMU: 1500 m, *Holttum SFN 25431* (BM, K, SING).

15.3.19. Ctenopteris subsecundodissecta (Zoll.) Copel., Philipp. J. Sci. 81: 114 (1953).

Epiphyte. Lower montane forest. Elevation: 1700–2400 m.

Material examined: KINATEKI RIVER HEAD: 2400 m, *Clemens 31789* (SING); PENATARAN BASIN: 1700 m, *Clemens 40142* (K, SING); SUMMIT TRAIL: 2300 m, *Parris & Croxall 8485* (K).

15.3.20. Ctenopteris taxodioides (Baker) Copel., Gen. Fil., 219 (1947).

Polypodium taxodioides Baker, J. Bot. 17: 42 (1879). C. Chr. & Holttum, Gardens' Bull. 7: 304 (1934). Type: Mount Kinabalu: *Burbidge s.n.* (holotype K!; isotype BM!).

Terrestrial. Lower and upper montane forest. Elevation: 2300–3000 m. Not endemic to Mount Kinabalu.

Additional material examined: EASTERN SHOULDER: 3000 m, *RSNB 849* (K, SING); GOKING'S VALLEY: 2700 m, *Fuchs 21474* (K); GURULAU SPUR: 2400 m, *Clemens 50959* (K), 2400 m, *51126* (US); KINATEKI RIVER HEAD: 2400 m, *Clemens 31784* (K); MARAI PARAI: 2300 m, *Clemens 33224* (K); MARAI PARAI SPUR: *Clemens 11040* (BM, K); SUMMIT TRAIL: 2400 m, *Parris & Croxall 8493* (K).

15.3.21. Ctenopteris venulosa (Blume) Kunze, Bot. Zeit. 4: 425 (1846).

Polypodium venulosum Blume, Enum. Pl. Javae, 128 (1828). C. Chr. & Holttum, Gardens' Bull. 7: 303 (1934). *Prosaptia venulosa* (Blume) M. G. Price, Contr. Univ. Mich. Herb. 16: 197 (1987).

Epiphyte. Lower and upper montane forest. Elevation: 2100–3000 m.

Material examined: DACHANG: 2900 m, *Clemens 28441* (BM, K); EASTERN SHOULDER: 2900 m, *RSNB 799* (K, SING); EASTERN SHOULDER, CAMP 4: 2700 m, *RSNB 1147* (K); GOKING'S VALLEY: 2700 m, *Fuchs 21465* (K); GURULAU SPUR: 2400 m, *Clemens 50921* (K); JANET'S HALT/SHEILA'S PLATEAU: 2900 m, *Collenette 21544* (K); KEMBURONGOH: 2100 m, *Holttum SFN 25524b* (BM, K, SING); KEMBURONGOH/PAKA-PAKA CAVE: 2700 m, *Molesworth-Allen 3256* (K), 2700 m, *3257* (K); KILEMBUN RIVER: 2700 m, *Clemens 33738* (K); KINATEKI RIVER HEAD: 2100 m, *Clemens 31794* (K); LAYANG-LAYANG: 2600 m, *Parris 11597* (K); MARAI PARAI: 2400–3000 m, *Clemens 33123* (K); SUMMIT TRAIL: 2300 m, *Parris & Croxall 8487* (K), 2300 m, *9171* (K).

15.3.22. Ctenopteris sp. 1

Epiphyte? Lower montane forest. Elevation: 1700 m. Endemic to Mount Kinabalu.

Material examined: PINOSUK PLATEAU: 1700 m, *RSNB 1869* (K).

15.4. GRAMMITIS

Copeland, E. B. 1952. *Grammitis*. Philipp. J. Sci. 80: 93–276. Parris, B. S. 1983. A taxonomic revision of the genus *Grammitis* Swartz (Grammitidaceae: Filicales) in New Guinea. Blumea 29: 13–222. Parris, B. S. 1986. Grammitidaceae of Peninsular Malaysia and Singapore. Kew Bull. 41: 508–514. Parris, B. S., Jermy, A. C., Camus, J. M., & Paul, A. M. 1984. The Pteridophyta of Gunung Mulu National Park. *In* Studies on the Flora of Gunung Mulu National Park, Sarawak. Ed. A. C. Jermy: 199–200. Forest Dept., Kuching, Sarawak.

15.4.1. Grammitis bongoensis (Copel.) Copel., Philipp. J. Sci. 80: 271, f. 90 (1952).

Polypodium bongoense Copel., Philipp. J. Sci. 38: 153 (1929). C. Chr. & Holttum, Gardens' Bull. 7: 295 (1934) p.p.

Epiphyte. Lower montane forest. Elevation: 1500–2400 m.

Material examined: KEMBURONGOH/LUMU-LUMU: 1800–2400 m, *Clemens 27998* (K); LUMU-LUMU: 1800 m, *Holttum SFN 25737* (SING); MARAI PARAI: 1500 m, *Clemens s.n.* (SING).

15.4.2. Grammitis caespitosa Blume, Enum. Pl. Javae, 115 (1828).

Epiphyte. Lower montane forest. Elevation: 1800 m.

Material examined: POWER STATION: 1800 m, *Parris & Croxall 8529* (K).

15.4.3. Grammitis clemensiae (Copel.) Parris, Fern Gaz. 12: 118 (1980).

Oreogrammitis clemensiae Copel., Philipp. J. Sci. 12C: 64 (1917). C. Chr. & Holttum, Gardens' Bull. 7: 292 (1934). Type: Mount Kinabalu: *Clemens 10618 p.p.* (lectotype of Parris, Blumea 29: 182 (1983), MICH!; isolectotype BM!).

Lithophyte. Summit zone and low upper montane forest. Elevation: 3400–4000 m. Not endemic to Mount Kinabalu.

Additional material examined: LOW'S PEAK: 4000 m, *Holttum SFN 25484* (SING); MOUNT KINABALU: 4000 m, *Holttum SFN 25484* (K); PANAR LABAN: 3400 m, *Parris & Croxall 8515* (K), 3400 m, *8516* (K); SUMMIT TRAIL: 3400 m, *Parris & Croxall 8757* (K).

15.4.4. Grammitis congener Blume, Enum. Pl. Javae, 115 (1828).

Polypodium congenerum (Blume) C. Presl, Tent., 180 (1836). C. Chr. & Holttum, Gardens' Bull. 7: 296 (1934). *Grammitis petrophila* Copel., Philipp. J. Sci. 56: 478, pl. 9 (1935). Type: Tinekuk River: 1700 m, *Clemens 50133* (lectotype of Parris, Kew Bull. 41: 511 (1986), MICH!; isolectotypes BM!, K!, UC!).

Lithophyte, epiphyte. Lower montane forest. Elevation: 1100–2700 m.

Additional material examined: KIAU VIEW TRAIL: 1600 m, *Parris & Croxall 9020* (K), 1600 m, *9118* (K); KILEMBUN BASIN: 1100 m, *Clemens 32570* (SING); KINATEKI RIVER: 2400 m, *Clemens 31729* (SING); KINATEKI RIVER HEAD: 2400 m, *Clemens 31729* (SING); LIWAGU RIVER TRAIL: 1400 m, *Edwards 2215* (K), 1500 m, *Parris 11493* (K), 1400 m, *Parris & Croxall 8480* (K), 1400 m, *9033* (K), 1500 m, *9064* (K), 1400 m, *9168* (K); MARAI PARAI: 1400 m, *Clemens 32387* (SING); MENTEKI RIVER: 1600 m, *Beaman 10779* (K, MICH, MSC); MESILAU BASIN: 2700 m, *Clemens 29061* (BM, K); MESILAU CAMP: *Poore 428* (K); MESILAU CAVE: 1900–2200 m, *Beaman 9566* (MICH); MESILAU CAVE TRAIL: 1700–1900 m, *Beaman 7979* (MICH, MSC), 1700–1900 m, *9106* (K, MICH, MSC); MOUNT KINABALU: 1500 m, *Shim SAN 75421* (SAN); PENATARAN BASIN: 1100 m, *Clemens 32570* (K, SING); SUMMIT TRAIL: 2400 m, *Parris & Croxall 8642* (K), 1800 m, *8652* (K); TENOMPOK: 1500 m, *Clemens 27792* (BM, SING), 1400 m, *Holttum SFN 25742* (BM, SING); WEST MESILAU RIVER: 1600–1700 m, *Beaman 8662* (K, MICH, MSC).

15.4.5. Grammitis dolichosora (Copel.) Copel., Philipp. J. Sci. 80: 188, f. 57 (1952).

Polypodium dolichosorum Copel., Philipp. J. Sci. 1, Suppl.: 159, pl. 16 (1906). C. Chr. & Holttum, Gardens' Bull. 7: 294 (1934).

Lithophyte. Lower montane forest, apparently rarely upper montane. Elevation: 1100–3500 m.

Material examined: KEMBURONGOH: 2100 m, *Holttum SFN 25533* (BM, K, SING); KILEMBUN BASIN: 1100 m, *Clemens s.n.* (SING), 1400 m, *33965* (K, SING), 1400 m, *34394* (K); KILEMBUN RIVER HEAD: 1400 m, *Clemens 32759* (BO); LIWAGU RIVER TRAIL: 1500 m, *Parris & Croxall 9063* (K); MARAI PARAI: 1500 m, *Clemens 32407* (K, SING), 1500 m, *32416* (K), 1500 m, *33101* (SING), 1500 m, *33103* (K); MT. NUNGKEK: 1100–1400 m, *Clemens 32759* (SING); PARK HEADQUARTERS: 1500 m, *Parris & Croxall 8524* (K); SUMMIT TRAIL: 3000–3500 m, *Jacobs 5758* (K).

15.4.6. Grammitis fasciata Blume, Enum. Pl. Javae, 116 (1828).

Polypodium fasciatum (Blume) C. Presl, Tent., 180 (1836). C. Chr. & Holttum, Gardens' Bull. 7: 293 (1934).

Lithophyte and low epiphyte. Lower and upper montane forest. Elevation: 1500–3400 m.

Material examined: EASTERN SHOULDER: 2900 m, *RSNB 836* (K, SING); EASTERN SHOULDER, CAMP 4: 2700 m, *RSNB 1146a* (K, SING); GURULAU SPUR: 3000 m, *Clemens 50909* (K); KEMBURONGOH/PAKA-PAKA CAVE: 3000 m, *Clemens 27059* (BM, SING), 2400–3400 m, *27999* (BM, K); KILEMBUN BASIN: 2700 m, *Clemens 33739* (BO, K, SING), 2700 m, *33742* (BO); LUBANG/PAKA-PAKA CAVE: *Clemens 10717* (BO); MARAI PARAI: 2400–3000 m, *Clemens 33124* (K), 2400–3000 m, *33210* (BO); MOUNT KINABALU: 1800–2400 m, *Clemens 30523* (BO); PAKA-PAKA

CAVE: *Clemens 27050* (BM, K), 3400 m, *28987* (BM, K), 2600–2700 m, *30462* (K), 3000 m, *Gibbs 4265* (BM), 2700 m, *Holttum SFN 25477* (BM, K, SING), 3000 m, *25870* (BM, K, SING), 2700 m, *Meijer SAN 22079* (K), *Topping 1663* (SING); PANAR LABAN: 3300 m, *Parris 11589* (K); SUMMIT TRAIL: 3000 m, *Edwards 2190* (K), *2193B* (K), 2500–2900 m, *Jacobs 5716* (K), 2400 m, *Parris & Croxall 8490* (K), 3000 m, *8510* (K); TENOMPOK/KEMBURONGOH: 1500–2100 m, *Clemens 27896* (BM).

15.4.7. Grammitis friderici-et-pauli (Christ) Copel., Philipp. J. Sci. 80: 250, f. 102 (1952).

Polypodium calcipunctatum Copel., Philipp. J. Sci. 12C: 61 (1917). C. Chr. & Holttum, Gardens' Bull. 7: 297 (1934). Type: Kemburongoh: *Clemens 10530* (lectotype of Parris, here designated, MICH!; isolectotype BM!).

Epiphyte. Lower montane forest. Elevation: 1700–2700 m.

Additional material examined: GOKING'S VALLEY: 2700 m, *Fuchs 21466* (K); GOLF COURSE SITE: 1700–2000 m, *Beaman 9613* (K, MICH, MSC); KEMBURONGOH: 2100 m, *Gibbs 4141* (BM), 2100 m, *Holttum SFN 25532* (BM, K, SING); MESILAU BASIN: *Clemens s.n.* (BM); SUMMIT TRAIL: 2300 m, *Parris & Croxall 8486* (K).

15.4.8. Grammitis havilandii (Baker) Copel., Gen. Fil., 211 (1947).

Polypodium havilandii Baker, Trans. Linn. Soc. Bot. 4: 253 (1894). C. Chr. & Holttum, Gardens' Bull. 7: 296 (1934). Type: Mount Kinabalu: 3200 m, *Haviland 1488* (holotype K!). *Polypodium multisorum* Copel., Philipp. J. Sci. 12C: 61 (1917). Type: Paka-paka Cave: *Topping 1665* (lectotype of Parris, Hook. Ic. Pl. 49, t. 3971 (1990), MICH!; isolectotype BM!).

Epiphyte. Lower and upper montane forest. Elevation: 1500–3200 m. Endemic to Mount Kinabalu.

Additional material examined: EASTERN SHOULDER: 3000 m, *RSNB 772* (K, SING), 3200 m, *Kitayama 64 UKMS 3817* (SAN); EASTERN SHOULDER, CAMP 4: 2700 m, *RSNB 1148* (K, SING), 2900 m, *1153* (K); KEMBURONGOH: 2400 m, *Clemens 27995* (K); KEMBURONGOH/PAKA-PAKA CAVE: 2700 m, *Holttum SFN 25480* (BM, SING), 2600 m, *Sinclair et al. 9107* (K, SING); MARAI PARAI: 1500 m, *Clemens s.n.* (SING); PAKA-PAKA CAVE: 2700 m, *Holttum SFN 25478* (BM, K, SING), 3100 m, *Parris 11664* (K), 3000 m, *Parris & Croxall 8715* (K), 3000 m, *8782* (K), *Topping 1668* (SING, US), *1711* (SING, US); PARK HEADQUARTERS: 1500 m, *Parris & Croxall 8468* (K); SUMMIT TRAIL: 3000 m, *Edwards 2193A* (K), 2500–2900 m, *Jacobs 5713* (K), 2800–2900 m, *5739* (K), *5797* (K), 2700 m, *Parris & Croxall 8512* (K).

15.4.9. Grammitis holttumii Copel., Univ. Calif. Publ. Bot. 18: 224 (1942).

Epiphyte. Lower to upper montane forest. Elevation: 1200–3000 m.

Material examined: LIWAGU RIVER TRAIL: 1700 m, *Parris & Croxall 9073* (K); PAKA-PAKA CAVE: 3000 m, *Parris & Croxall 8721* (K); PENIBUKAN: 1200 m, *Clemens 32142* (SING); POWER STATION: 1800 m, *Parris & Croxall 9066* (K); SUMMIT TRAIL: 2600 m, *Parris 11557* (K), 2600 m, *Parris & Croxall 8502* (K), 2600 m, *8660* (K), 2700 m, *8722* (K), 2900 m, *8788* (K).

15.4.10. Grammitis impressa Copel., Philipp. J. Sci. 80: 242, f. 95 (1952).

Epiphyte. Lower montane forest. Elevation: 1100–1400 m.

Material examined: MT. NUNGKEK: 1100–1400 m, *Clemens 32759A* (SING).

15.4.11. Grammitis intromissa (Christ) Parris, Fern Gaz. 12: 180 (1981).

Polypodium setigerum Blume, Enum. Pl. Javae, 123 (1828). C. Chr. & Holttum, Gardens' Bull. 7: 296 (1934).

Epiphyte. Lower montane forest, usually growing on *Cyathea* spp. Elevation: 1400–2400 m.

Material examined: KEMBURONGOH: 2100 m, *Holttum SFN 25739* (BM, SING); KILEMBUN RIVER HEAD: 1400 m, *Clemens 32463* (BO); SUMMIT TRAIL: 2400 m, *Parris & Croxall 8492* (K), 2400 m, *8851* (K).

15.4.12. Grammitis jagoriana (Mett. ex Kuhn) Tagawa, Acta Phytotax. Geobot. 10: 284 (1941).

Grammitis cuneifolia Copel., Philipp. J. Sci. 56: 479, pl. 8, f. 2 (1935). Type: Penibukan: 2100 m, *Clemens 40962 p.p.* (lectotype of Parris, here designated, MICH!).

Epiphyte. Lower montane forest and low forest on ultramafics. Elevation: 1400–2600 m.

Additional material examined: KILEMBUN BASIN: 1400 m, *Clemens 40016* (K, SING); LIWAGU RIVER TRAIL: 1500 m, *Parris & Croxall 8483* (K); SUMMIT TRAIL: 2600 m, *Parris & Croxall 8658* (K).

15.4.13. Grammitis kinabaluensis (Copel.) Copel., Philipp. J. Sci. 56: 479 (1935).

Polypodium kinabaluense Copel., Philipp. J. Sci. 56: 479 (1935). C. Chr. & Holttum, Gardens' Bull. 7: 292 (1934). Type: Summit Trail: *Clemens 10649* (lectotype of Parris, here designated, MICH!; isolectotypes BM!, UC!). *Polypodium malaicum* sensu C. Chr. & Holttum non Alderw., Gardens' Bull. 7: 292 (1934).

Epiphyte and lithophyte. Rarely lower montane forest, mostly upper montane forest and summit zone. Elevation: 1500–4000 m. Endemic to Mount Kinabalu.

Additional material examined: EASTERN SHOULDER: 3100 m, *RSNB 851* (K, SING); GURULAU SPUR: 3000 m, *Clemens 51110* (K); KEMBURONGOH/PAKA-PAKA CAVE: 2700 m, *Holttum SFN 25479* (BM, K, SING); KILEMBUN BASIN: 2700 m, *Clemens s.n.* (K); KILEMBUN RIVER: 2700 m, *Clemens 33789* (SING); KILEMBUN RIVER HEAD: 2700 m, *Clemens 33927A* (BO); MARAI PARAI: 3200 m, *Clemens 32347* (K, SING); MARAI PARAI SPUR: 1800 m, *Holttum SFN 25613* (SING); MINIRINTEG (MESILAU BASIN): 1500 m, *Clemens s.n.* (LAE); MOUNT KINABALU: 3700–4000 m, *Holttum SFN 25483* (BM, K, SING), 3400 m, *Kitayama K742 UKMS 3903* (SAN), 2700 m, *Meijer SAN 22032* (SAN); PAKA-PAKA CAVE: *Clemens s.n.* (BM), 3700–4000 m, *27783* (K), 2700 m, *27990* (US), 3000 m, *Holttum SFN 25744* (BM, SING), 3000 m, *25745* (SING), 3000 m, *25873* (BM, SING), 3100 m, *Parris 11665* (K), 3000 m, *Parris & Croxall 8701* (K), 3000 m, *8709* (K); PAKA-PAKA CAVE/LUMU-LUMU: 1800–3400 m, *Clemens 28981* (BM); PENATARAN BASIN: 2700 m, *Clemens 33666* (SING); SAYAT-SAYAT: 3700 m, *Parris & Croxall 8760* (K); SUMMIT AREA: 3400 m, *Clemens s.n.* (BM), 3700–4000 m, *51111* (K), 3300 m, *Sinclair et al. 9152* (K, SING); SUMMIT TRAIL: 3000 m, *Edwards 2204* (K), 2400 m, *Edwards 2219* (K), 2500–2900 m, *Jacobs 5735* (K), 2900 m, *5740* (K), *5794* (K), 2700 m, *Parris & Croxall 8513* (K), 3200 m, *8753* (K), 3300 m, *8765* (K).

15.4.14. Grammitis knutsfordiana (Baker) Copel., Univ. Calif. Publ. Bot. 18: 224 (1942).

Polypodium warburgii Christ, Monsunia 1: 59 (1900). C. Chr. & Holttum, Gardens' Bull. 7: 295 (1934).

Epiphyte. Low on trees in lower and upper montane forest. Elevation: 900–3000 m.

Material examined: EASTERN SHOULDER, CAMP 4: 2700 m, *RSNB 1146b* (K, SING), 2900 m, *1159c* (K); GURULAU SPUR: 2400 m, *Clemens 51190* (UC); KEMBURONGOH: *Clemens s.n.* (BM), 2100 m, *Holttum SFN 25741* (BM, K, SING); KINATEKI RIVER: 2700 m, *Clemens 31921* (SING); KINATEKI RIVER HEAD: 2700 m, *Clemens 31921* (SING); LIWAGU RIVER TRAIL: 1700 m, *Parris & Croxall 9067* (K), 1800 m, *9071* (K); LUMU-LUMU: 2300 m, *Clemens 27991* (BM); MARAI PARAI: *Clemens 11064* (BM); MESILAU BASIN: 900–2400 m, *Clemens 29035* (BM, SING); PAKA-PAKA CAVE: 2700–3000 m, *Gibbs 4255* (BM); SUMMIT TRAIL: 2400 m, *Jacobs 5796* (K), 2500 m, *Parris & Croxall 8491* (K), 2500 m, *8825* (K).

15.4.15. Grammitis locellata (Baker) Copel., Occ. Papers Bishop Mus. 15: 86 (1939).

Polypodium caespitosum sensu C. Chr. & Holttum non (Blume) Mett., Gardens' Bull. 7: 292 (1934).

Epiphyte and lithophyte. Short upper montane forest and in summit zone in rock crevices. Elevation: 3300–3700 m.

Material examined: MOUNT KINABALU: 3700 m, *Holttum SFN 25496* (BM, SING); PANAR LABAN: 3300 m, *Parris 11591* (K); SAYAT-SAYAT: 3700 m, *Parris & Croxall 8759* (K); SUMMIT TRAIL: 3300 m, *Parris & Croxall 8514* (K), 3300 m, *8762* (K).

15.4.16. Grammitis oblanceolata (Baker) Copel., Philipp. J. Sci. 80: 166, f. 34 (1952).

Epiphyte. Lower montane valley forest. Elevation: 1100–1500 m.

Material examined: LIWAGU RIVER TRAIL: 1400 m, *Parris & Croxall 9169* (K); MT. NUNGKEK: 1100–1400 m, *Clemens 32759* (SING); PARK HEADQUARTERS: 1500 m, *Parris & Croxall 8526* (K).

15.4.17. Grammitis padangensis (Baker) Copel., Philipp. J. Sci. 80: 212, f. 75 (1952).

Polypodium padangense Baker, J. Bot. 18: 213 (1880). C. Chr. & Holttum, Gardens' Bull. 7: 293 (1934).

Epiphyte and lithophyte. Upper montane short and tall forest and summit area. Elevation: 1800–3800 m.

Material examined: KEMBURONGOH/PAKA-PAKA CAVE: 2800–2900 m, *Jacobs 5738* (K); MESILAU BASIN: 3400 m, *Smith 533* (L); PAKA-PAKA CAVE: 3000 m, *Clemens 27997* (K, SING), 2700–3000 m, *Gibbs 4272* (BM), 3100 m, *Parris 11666* (K); PAKA-PAKA CAVE/LUMU-LUMU: 1800–3400 m, *Clemens 28981A* (BM); PANAR LABAN: 3300 m, *Parris 11585* (K); SUMMIT AREA: 3800 m, *Collenette 615* (K); SUMMIT TRAIL: 3000 m, *Parris & Croxall 8503* (K), 3200 m, *8750* (K).

15.4.18. Grammitis pilosiuscula Blume, Enum. Pl. Javae, 115 (1828).

Epiphyte. Lower montane forest. Elevation: 1600–1700 m.

Material examined: KIAU VIEW TRAIL: 1600 m, *Parris & Croxall 8522* (K), 1600 m, *9117* (K); LIWAGU RIVER TRAIL: 1700 m, *Parris & Croxall 9072* (K); PARK HEADQUARTERS: 1600 m, *Parris & Croxall 9019* (K).

15.4.19. Grammitis reinwardtii Blume, Enum. Pl. Javae Addend., 2 (1828).

Epiphyte, usually low on trees. Lower montane valley forest. Elevation: 900–1800 m.

Material examined: MARAI PARAI: 1500 m, *Clemens s.n.* (BO), 1800 m, *Gibbs 4062* (BM); MARAI PARAI SPUR: 1500 m, *Parris 11662* (K); MOUNT KINABALU: *Clemens 11047* (BM), *Shim SAN 75402* (SAN); MT. NUNGKEK: 1100–1400 m, *Clemens 32759* (BO); PARK HEADQUARTERS: 1500 m, *Parris 11434* (K); TAHUBANG RIVER: 900–1100 m, *Kanis SAN 51471* (SAN).

15.4.20. Grammitis reinwardtioides Copel., Philipp. J. Sci. 56: 478, pl. 10 (1935). Type: Tahubang River: 1200 m, *Clemens 40792* (lectotype of Parris, here designated, MICH!; isolectotype BM!).

Polypodium reinwardtii sensu C. Chr. & Holttum non (Blume) C. Presl, Gardens' Bull. 7: 295 (1934). *Polypodium reinwardtii* var. an sp. nov.? C. Chr. in C. Chr. & Holttum, Gardens' Bull. 7: 295 (1934).

Epiphyte and lithophyte. Lower montane valley forest. Elevation: 1400–2400 m. Not endemic to Mount Kinabalu.

Additional material examined: GOLF COURSE SITE: 1700 m, *Beaman 8772* (MICH); KEMBURONGOH: *Clemens s.n.* (BM), 2100 m, *Holttum SFN 25738* (BM, K, SING); KEMBURONGOH/ LUMU-LUMU: 1800–2400 m, *Clemens 28968A* (BM), 1800 m, *Holttum SFN 25737* (BM, SING); KILEMBUN RIVER: 1400 m, *Clemens 32463* (K, SING); LIWAGU RIVER TRAIL: 1500 m, *Parris & Croxall 9065* (K); LUMU-LUMU: 1800 m, *Holttum SFN 25737* (BM); MARAI PARAI: 1500 m, *Clemens s.n.* (SING); PARK HEADQUARTERS: 1400 m, *Edwards 2166* (K), 1500 m, *Parris 11433* (K), *Parris & Croxall 8477* (K); SUMMIT TRAIL: 2300 m, *Parris & Croxall 8634* (K); TENOMPOK: 1500 m, *Clemens s.n.* (BM), 1500 m, *27792* (BM), 1400 m, *Holttum SFN 25399* (BM, K); TENOMPOK/LUMU-LUMU: 1500 m, *Holttum SFN 25743* (BM, K, SING); TENOMPOK/TOMIS: *Clemens s.n.* (BM); WEST MESILAU RIVER: 1600–1700 m, *Beaman 8697* (MICH, MSC).

15.4.21. Grammitis scabristipes (Baker) Copel., Univ. Calif. Publ. Bot. 18: 223 (1942).

Polypodium obscurum sensu C. Chr. & Holttum non (Blume) Mett., Gardens' Bull. 7: 294 (1934).

Epiphyte. Lower montane to tall upper montane forest. Elevation: 1800–3000 m.

Material examined: MESILAU RIVER: 1800 m, *Clemens 51367* (K); PAKA-PAKA CAVE: 3000 m, *Holttum SFN 25871* (BM, K, SING); SUMMIT TRAIL: 3000 m, *Parris & Croxall 8509* (K).

15.4.22. Grammitis sumatrana (Baker) Copel., Philipp. J. Sci. 56: 105 (1935).

Polypodium sumatranum Baker, J. Bot. 18: 214 (1880). C. Chr. & Holttum, Gardens' Bull. 7: 297 (1934).

Epiphyte and lithophyte. Lower montane to tall upper montane forest. Elevation: 1500–3400 m.

Material examined: BAMBANGAN RIVER: 3000 m, *Kitayama K716 UKMS 3877* (SAN); EAST MESILAU RIVER: 2100 m, *Clemens 28240* (BM); EASTERN SHOULDER: 2900 m, *RSNB 829* (K, SING); EASTERN SHOULDER, CAMP 4: 2900 m, *RSNB 1159b* (K); GOLF COURSE SITE: 1700–2000 m, *Beaman 9613a* (MICH); GURULAU SPUR: *Clemens 50911* (US); KILEMBUN RIVER: 2400 m, *Clemens 33743* (K); MARAI PARAI: *Clemens 10920* (BM), 1500 m, *32407* (SING); PAKA-PAKA CAVE: 3400 m, *Clemens 27065* (US), 3000 m, *Holttum SFN 25503* (BM, K, SING); SUMMIT TRAIL: 2800 2900 m, *Jacobs 5738* (K), *5793* (K), 3000 m, *Parris & Croxall 8511* (K), 2400 m, *8643* (K).

15.4.23. Grammitis sp. 1

Epiphyte. Lower montane forest. Elevation: 1500 m. Not endemic to Mount Kinabalu.

Material examined: SEDIKEN RIVER: 1500 m, *Clemens 32287* (BO).

15.4.24. Grammitis sp. 2

Epiphyte. Lower montane forest. Elevation: 1500 m. Endemic to Mount Kinabalu.

Material examined: LIWAGU RIVER TRAIL: 1500 m, *Parris & Croxall 8482* (K), 1500 m, *9040* (K); MEMPENING TRAIL: 1500 m, *Parris & Croxall 9115* (K).

15.4.25. Grammitis sp. 3

Epiphyte. Low upper montane forest on ultramafics. Elevation: 1500–2600 m. Endemic to Mount Kinabalu.

Material examined: MARAI PARAI SPUR: 1500 m, *Parris 11526* (K); SUMMIT TRAIL: 2600 m, *Parris & Croxall 8501* (K), 2600 m, *8659* (K).

15.4.26. Grammitis sp. 4

Epiphyte. Low upper montane forest. Elevation: 3400 m. Endemic to Mount Kinabalu.

Material examined: PANAR LABAN: 3400 m, *Parris & Croxall 8525* (K).

15.4.27. Grammitis sp. 5

Lithophyte. Low upper montane forest. Elevation: 3300–3400 m. Endemic to Mount Kinabalu.

Material examined: PANAR LABAN: 3400 m, *Parris & Croxall 8763* (K); SUMMIT TRAIL: 3300 m, *Parris & Croxall 8530* (K), 3400 m, *8758* (K).

15.4.28. Grammitis sp. 6

Lithophyte and epiphyte. Upper montane forest. Elevation: 3000–3300 m. Endemic to Mount Kinabalu.

Material examined: SUMMIT TRAIL: 3000 m, *Parris & Croxall 8531* (K), 3300 m, *8744* (K), 3200 m, *8751* (K).

15.4.29. Grammitis sp. 7

Polypodium lasiosorum sensu C. Chr. & Holttum non (Blume) Hook., Gardens' Bull. 7: 294 (1934).

Epiphyte. Upper montane forest. Elevation: 1800–2900 m. Endemic to Mount Kinabalu.

Material examined: KEMBURONGOH/LUMU-LUMU: 1800–2400 m, *Clemens 27998* (K); KEMBURONGOH/PAKA-PAKA CAVE: 2700 m, *Holttum SFN 25872* (BM, K, SING); MARAI PARAI: 2700 m, *Clemens s.n.* (SING); SUMMIT TRAIL: 2400 m, *Edwards 2218* (K), 2500–2900 m, *Jacobs 5712* (K), 2600 m, *Parris 11667* (K), 2600 m, *Parris & Croxall 8500* (K), 2600 m, *8647* (K), 2600 m, *8657* (K), 2700 m, *8723* (K), 2800 m, *8836* (K).

15.5. PROSAPTIA

Parris, B. S. 1986. Grammitidaceae of Peninsular Malaysia and Singapore. Kew Bull. 41: 500–502 [as Ctenopteris].

15.5.1. Prosaptia alata (Blume) Christ, Ann. Jard. Buit. 2, 5: 127 (1905).

Polypodium emersonii (Hook. & Grev.) C. Chr. in C. Chr. & Holttum, Gardens' Bull. 7: 303 (1934).

Epiphyte and terrestrial. Montane dipterocarp forest. Elevation: 800–1300 m.

Material examined: DALLAS: 800 m, *Holttum SFN 25366* (K, SING); LUGAS HILL: 1300 m, *Beaman 10543* (MICH, MSC); PENIBUKAN: 1200 m, *Clemens s.n.* (SING); SAYAP: 800–1000 m, *Beaman 9810* (MICH).

15.5.2. Prosaptia contigua (G. Forster) C. Presl, Tent., 166 (1836).

Polypodium contiguum (G. Forster) J. Smith, J. Bot. 3: 394 (1841). C. Chr. & Holttum, Gardens' Bull. 7: 303 (1934).

Epiphyte. Lower montane to upper montane tall forest. Elevation: 1200–2900 m.

Material examined: DACHANG: 2900 m, *Clemens 28317* (BM, K); EASTERN SHOULDER, CAMP

4: 2700 m, *RSNB 1149* (K); KILEMBUN RIVER: 2300 m, *Clemens 33716* (US); KUNDASANG: 1200 m, *RSNB 1440* (K); LIWAGU RIVER TRAIL: 1400 m, *Edwards 2175* (K); LUMU-LUMU: 1800 m, *Holttum s.n.* (BM, SING); MAMUT COPPER MINE: 1600–1700 m, *Beaman 9943a* (MICH); MARAI PARAI SPUR: *Clemens 11034* (UC); PARK HEADQUARTERS: 1500 m, *Parris & Croxall 8473* (K); PENATARAN BASIN: 2000 m, *Clemens 40156* (UC); PINOSUK PLATEAU: 1500 m, *Parris & Croxall 9137* (K); SUMMIT TRAIL: 2400 m, *Parris & Croxall 8645* (K); TENOMPOK: 1500 m, *Clemens 26981* (BM, K), 1500 m, *28681* (BM, K); TENOMPOK/LUMU-LUMU: 1500 m, *Holttum s.n.* (SING); WEST MESILAU RIVER: 1600–1700 m, *Beaman 8677* (MICH).

15.5.3. Prosaptia davalliacea (F. Muell. & Baker) Copel., Univ. Calif. Publ. Bot. 12: 404 (1931).

Polypodium davalliaceum F. Muell. & Baker, J. Bot. 28: 108 (1890). C. Chr. & Holttum, Gardens' Bull. 7: 303 (1934).

Epiphyte and lithophyte. Tall and short upper montane forest and summit area. Elevation: 2400–3800 m.

Material examined: GURULAU SPUR: 3400–3700 m, *Clemens 50900* (K); KEMBURONGOH: 2400–2700 m, *Clemens 27993* (K); MARAI PARAI: 2900 m, *Clemens 33206* (K, SING); MOUNT KINABALU: 3200 m, *Haviland 1484* (K), 2700 m, *Meijer SAN 22062* (K, SAN); PAKA-PAKA CAVE: 3000 m, *Holttum SFN 25506* (BM, K, SING), 3000 m, *Meijer SAN 29272* (K), 3100 m, *Parris 11581* (K), *Topping 1679* (SING), *1715* (BM, SING); SUMMIT AREA: 3800 m, *Collenette 618* (K); SUMMIT TRAIL: 2700–2800 m, *Jacobs 5742* (K), 2400–2900 m, *5767* (K), 3000 m, *Parris & Croxall 8505* (K).

15.6. SCLEROGLOSSUM

Parris, B. S. 1986. Grammitidaceae of Peninsular Malaysia and Singapore. Kew Bull. 41: 514–516.

15.6.1. Scleroglossum debile (Mett. ex Kuhn) Alderw., Bull. Jard. Bot. Buit. 2, 7: 39 (1912). C. Chr. & Holttum, Gardens' Bull. 7: 291 (1934).

Epiphyte. Lower montane forest. Elevation: 1800 m.

Material examined: MARAI PARAI SPUR: 1800 m, *Holttum SFN 25613* (BM, K).

15.6.2. Scleroglossum minus (Fée) C. Chr. agg.

a. (large form)

Epiphyte and lithophyte. Tall and short upper montane forest, particularly on ultramafics. Elevation: 2600–3200 m. Not endemic to Mount Kinabalu.

Material examined: LAYANG-LAYANG: 2600 m, *Parris & Croxall 8754* (K); SUMMIT TRAIL: 3000 m, *Edwards 2192* (K), 3000 m, *Parris 11567* (K), 2800 m, *Parris & Croxall 8519* (K), 3200 m, *8755* (K), 2900 m, *8789* (K).

b. (narrow form)

Scleroglossum minus sensu C. Chr. & Holttum, Gardens' Bull. 7: 292 (1934). *Scleroglossum pusillum* (Blume) Alderw. var. *angustissimum* sensu C.Chr. & Holttum non (Copel.) C.Chr., Gardens' Bull. 7: 292 (1934). *Scleroglossum sulcatum* sensu C.Chr. & Holttum non (Kuhn) Alderw. Gardens' Bull. 7: 292 (1934).

Epiphyte. Lower and upper montane forest. Elevation: 1500–3700 m. Not endemic to Mount Kinabalu.

Material examined: KADAMAIAN RIVER HEAD: 3400 m, *Clemens 50906* (K); KEMBURONGOH: *Clemens s.n.* (BM), 2100 m, *Holttum SFN 25542* (K, SING); KEMBURONGOH/PAKA-PAKA CAVE: 2100–2700 m, *Molesworth-Allen 3259* (K); KIAU VIEW TRAIL: 1700 m, *Parris & Croxall 8565* (K); KILEMBUN RIVER: 2900 m, *Clemens 51368* (K), 2700–3000 m, *51368a* (K); LIWAGU RIVER TRAIL: 1600 m, *Parris & Croxall 9074* (K); MEMPENING TRAIL: 1600 m, *Parris 11660* (K); MT. NUNGKEK: 1700 m, *Clemens 32811* (K); PAKA-PAKA CAVE: 3700 m, *Clemens 27998* (K), 3200 m, *Haviland 1481* (K);

PARK HEADQUARTERS: 1500 m, *Parris & Croxall 8475* (K); SUMMIT TRAIL: 2500–2900 m, *Jacobs 5721* (K), 1900 m, *Parris & Croxall 8654* (K), 2100 m, *8655* (K).

15.6.3. Scleroglossum aff. minus (Fée) C. Chr.

Epiphyte. Lowland dipterocarp forest. Elevation: 1000 m. Not endemic to Mount Kinabalu.

Material examined: PORING HOT SPRINGS/LANGANAN WATER FALLS: 1000 m, *Parris & Croxall 8935* (K), 1000 m, *8943* (K).

15.6.4. Scleroglossum pusillum (Blume) Alderw., Bull. Jard. Bot. Buit. 2, 7: 39, t. 5, f. 1, 2 (1912). C. Chr. & Holttum, Gardens' Bull. 7: 291 (1934).

Scleroglossum angustissimum Copel., Philipp. J. Sci. 12C: 65 (1917). Type: Marai Parai Spur: *Clemens 11048* (lectotype of Parris, here designated, MICH!).

Epiphyte. Lower and upper montane forest. Elevation: 1500–3000 m.

Additional material examined: EASTERN SHOULDER, CAMP 4: 2900 m, *RSNB 1160* (K, SING); GURULAU SPUR: 1500–1800 m, *Clemens 50583* (K); KEMBURONGOH/LUMU-LUMU: 1800 m, *Holttum SFN 25464* (BM, K, SING); KEMBURONGOH/PAKA-PAKA CAVE: 2400–3000 m, *Gibbs 4068* (BM); LIWAGU RIVER TRAIL: 1600 m, *Parris & Croxall 9068* (K); LUBANG/PAKA-PAKA CAVE: *Clemens 10722* (BM); MARAI PARAI: 1500 m, *Clemens 33160* (K), 1500 m, *Holttum SFN 25179* (BM, K, SING); MARAI PARAI SPUR: 3000 m, *Clemens 32348* (SING), 1500 m, *Parris 11534* (K); PAKA-PAKA CAVE: 3000 m, *Holttum SFN 25736* (K, SING); PAKA-PAKA CAVE/LUMU-LUMU: 1800–3000 m, *Clemens 28982* (BM), 2700 m, *29104* (BM); PARK HEADQUARTERS: 1500 m, *Parris & Croxall 9021* (K); SUMMIT TRAIL: 2700 m, *Parris 11558* (K), 2500 m, *Parris & Croxall 8489* (K), 2600 m, *8648* (K), 2800 m, *8725* (K).

15.7. XIPHOPTERIS

Parris. 1986. Grammitidaceae of Peninsular Malaysia and Singapore. Kew Bull. 41: 502–505.

15.7.1. Xiphopteris cornigera (Baker) Copel., Gen. Fil., 215 (1947).

Polypodium hieronymusii sensu C. Chr. & Holttum p.p., Gardens' Bull. 7: 300 (1934) non C. Chr. *Polypodium minutum* sensu C. Chr. & Holttum non Blume, Gardens' Bull. 7: 300 (1934). *Polypodium hecistophyllum* Copel., Philipp. J. Sci. 56: 477, t. 8, f. 1 (1935). Type: Penibukan: 1200 m, *Clemens 40837* (lectotype of Parris, here designated, MICH!; isolectotypes BM!, K!, UC!).

Epiphyte and lithophyte. Lower montane valley forest. Elevation: 1200–1600 m.

Additional material examined: GURULAU SPUR: *Gibbs 4018* (BM, K); LIWAGU RIVER: 1400 m, *Parris & Croxall 9174* (K); LIWAGU RIVER TRAIL: 1400 m, *Edwards 2176C* (K), *2218* (K), *2229* (K), 1400 m, *Parris & Croxall 9172* (K), 1500 m, *9173* (K); MEMPENING TRAIL: 1600 m, *Parris 11499* (K); MOUNT KINABALU: 1500 m, *Shim SAN 75458* (SAN); PARK HEADQUARTERS: 1500 m, *Parris 10781* (K), 1500 m, *Parris & Croxall 8469* (K), 1500 m, *9024* (K); PENATARAN RIVER: 1300 m, *Clemens 34279* (K, SING), PENIBUKAN: 1200 m, *Clemens 30609* (K).

15.7.2. Xiphopteris musgraviana (Baker) Parris, Kew Bull. 41: 69 (1986).

Polypodium bryophyllum Alderw., Bull. Jard. Bot. Buit. 2, 16: 35 (1914). C. Chr. & Holttum, Gardens' Bull. 7: 299 (1934).

Epiphyte. Lower montane ridge forest. Elevation: 1500–1900 m.

Material examined: KEMBURONGOH/LUMU-LUMU: 1800 m, *Holttum SFN 25467* (BM, K, SING); LIWAGU RIVER TRAIL: 1400 m, *Edwards 2176D* (K); PARK HEADQUARTERS: 1500 m, *Parris & Croxall 8470* (K); SUMMIT TRAIL: 1900 m, *Parris & Croxall 8653* (K).

15.7.3. Xiphopteris nudicarpa (P. M. Zamora & Co) Parris, comb. nov.

Acrosorus nudicarpus P. M. Zamora & Co, Nat. Appl. Sci. Bull. 32: 47, t. 4 (1980). *Grammitis palasaba* M.G. Price, Contr. Univ. Mich. Herb. 16: 196 (1987). *Polypodium alternidens* sensu C. Chr. & Holttum non Ces., Gardens' Bull. 7: 299 (1934).

Terrestrial, epiphytic and lithophytic. Upper montane forest, apparently mostly on ultramafics. Elevation: 2400–3000 m.

Material examined: KEMBURONGOH: *Clemens 27989* (BM, K); KEMBURONGOH/PAKA-PAKA CAVE: 2400–3000 m, *Gibbs 4299* (BM), 2700 m, *Holttum SFN 25476* (BM, K); LAYANG-LAYANG: 2700 m, *Edwards 2195* (K); LUBANG/PAKA-PAKA CAVE: *Clemens 10721* (UC); MOUNT KINABALU: *Burbidge s.n.* (K); SUMMIT TRAIL: 2500–2900 m, *Jacobs 5718 (K).*

15.7.4. Xiphopteris subpinnatifida (Blume) Copel., Gen. Fil., 215 (1947).

Terrestrial. Lower montane forest. Elevation: 1500–2100 m.

Material examined: KEMBURONGOH: 2100 m, *Holttum SFN 25740* (SING); MARAI PARAI SPUR: 1500 m, *Clemens 11072* (BM).

16. HYMENOPHYLLACEAE

Copeland, E. B. 1938. Genera Hymenophyllacearum. Philipp. J. Sci. 67: 1–110.

16.1. CALLISTOPTERIS

Copeland, E. B. 1933. *Trichomanes*. Philipp. J. Sci. 51: 227–232. Copeland, E. B. 1938. Genera Hymenophyllacearum. Philipp. J. Sci. 67: 64–65.

16.1.1. Callistopteris apiifolia (C. Presl) Copel., Philipp. J. Sci. 67: 65 (1938).

Trichomanes apiifolium C. Presl, Hymen., 16, 44 (1843). C. Chr. & Holttum, Gardens' Bull. 7: 218 (1934).

Terrestrial. Lower montane forest. Elevation: 1300–1900 m.

Material examined: KILEMBUN RIVER: 1400 m, *Clemens 32513* (K, SING); KILEMBUN RIVER HEAD: 1500–1800 m, *Clemens 33984* (K); LIWAGU RIVER TRAIL: 1500 m, *Parris 11488* (K), 1500 m, *Parris & Croxall 9060* (K); LUMU-LUMU: 1900 m, *Sinclair et al. 9222* (K, SING); PARK HEADQUARTERS: 1500–1600 m, *Kokawa & Hotta 2894* (K); TENOMPOK: 1500 m, *Clemens 28225* (K), 1300 m, *Holttum SFN 25396* (K, SING); TINEKUK FALLS: 1800 m, *Clemens 40877* (K).

16.2. CEPHALOMANES

Copeland, E. B. 1933. *Trichomanes*. Philipp. J. Sci. 51: 245–257. Copeland, E. B. 1938. Genera Hymenophyllacearum. Philipp. J. Sci. 67: 66–68.

16.2.1. Cephalomanes javanicum (Blume) Bosch, Abh. Bohm. Ges. V. 5: 334 (1848).

Terrestrial. Lower montane forest. Elevation: 1200 m.

Material examined: MARAI PARAI: 1200 m, *Clemens 32486* (K).

16.2.2. Cephalomanes laciniatum (Roxb.) De Vol in H. L. Li et al., Fl. Taiwan 1: 98 (1975).

Terrestrial. Lowland dipterocarp forest to lower montane forest. Elevation: 500–1400 m.

Material examined: LUGAS HILL: 1300 m, *Beaman 10528* (MICH); PINOSUK PLATEAU: 1400 m, *Beaman 10728* (MICH); PORING HOT SPRINGS: 500 m, *Parris & Croxall 8993* (K); PORING HOT SPRINGS/LANGANAN WATER FALLS: 1000 m, *Parris & Croxall 8929* (K).

16.3. CREPIDOMANES

Copeland, E. B. 1933. *Trichomanes*. Philipp. J. Sci. 51: 174–196. Copeland, E. B. 1938. Genera Hymenophyllacearum. Philipp. J. Sci. 67: 58–61.

16.3.1. Crepidomanes bilabiatum (Nees & Blume) Copel., Philipp. J. Sci. 67: 59 (1938).

Trichomanes bilabiatum Nees & Blume, Nova Acta 11: 123, t. 13, f. 2 (1823). C. Chr. & Holttum, Gardens' Bull. 7: 216 (1934).

Epiphyte. Lower montane forest. Elevation: 1200–1500 m.

Material examined: KIAU/LUBANG: *Topping 1813* (SING); LUBANG: 1200 m, *Holttum SFN 25554* (K, SING); TENOMPOK: 1500 m, *Clemens 26858* (K).

16.3.2. Crepidomanes bipunctatum (Poir.) Copel., Philipp. J. Sci. 67: 59 (1938).

Trichomanes bipunctatum Poir., Encycl. 8: 69 (1808). C. Chr. & Holttum, Gardens' Bull. 7: 216 (1934).

Epiphyte. Lowland dipterocarp forest to lower montane forest. Elevation: 300–1800 m.

Material examined: BAT CAVE: 600 m, *Parris & Croxall 8972* (K); DALLAS/TENOMPOK: 1200 m, *Clemens 27588* (K); KEBAYAU: 300 m, *Holttum SFN 25627* (SING); KIAU/LUBANG: *Topping 1600* (SING), *1806* (SING); KIPUNGIT FALLS: 600 m, *Parris & Croxall 8919* (K); LINISIHANG RIVER: 300 m, *Holttum SFN 25627* (K); LIWAGU RIVER TRAIL: 1400 m, *Parris & Croxall 8620* (K), 1400 m, *9043* (K); PORING HOT SPRINGS: *Kokawa & Hotta 4836* (K), 500 m, *Parris & Croxall 8995* (K); SUMMIT TRAIL: 1800 m, *Parris & Croxall 8879* (K); TENOMPOK: 1500 m, *Clemens 28232* (K), 1500 m, *28302* (K), 1500 m, *28421* (SING), 1400 m, *Holttum SFN 25400* (K, SING); TIBABAR FALLS: 1800 m, *Parris 11464* (K).

16.3.3. Crepidomanes brevipes (C. Presl) Copel., Philipp. J. Sci. 67: 60 (1938).

Trichomanes brevipes (C. Presl) Baker, Syn. Fil., 84 (1867). C. Chr. & Holttum, Gardens' Bull. 7: 217 (1934).

Epiphyte. Hill forest and lower montane forest. Elevation: 900–1300 m.

Material examined: DALLAS: 1100 m, *Holttum SFN 25142* (K); EASTERN SHOULDER: 900 m, *RSNB 1598* (K, SING); LUGAS HILL: 1300 m, *Beaman 10557* (MICH).

16.3.4. Crepidomanes christii (Copel.) Copel., Philipp. J. Sci. 67: 60 (1938).

Epiphyte. Lowland dipterocarp forest. Elevation: 600 m.

Material examined: PORING HOT SPRINGS/LANGANAN WATER FALLS: 600 m, *Parris & Croxall 8971* (K).

16.4. GONOCORMUS

Copeland, E. B. 1933. *Trichomanes.* Philipp. J. Sci. 51: 143–153. Copeland, E. B. 1938. Genera Hymenophyllacearum. Philipp. J. Sci. 67: 56–57.

16.4.1. Gonocormus alagensis (Christ) Copel., Philipp. J. Sci. 67: 57 (1938).

Epiphyte. Lower montane forest. Elevation: 2500 m.

Material examined: SUMMIT TRAIL: 2500 m, *Parris & Croxall 8675* (K).

16.4.2. Gonocormus diffusus (Blume) Bosch, Hymen. Jav., 9, t. 4 (1861).

Epiphyte. Lower montane forest. Elevation: 1500–1600 m.

Material examined: KIAU VIEW TRAIL: 1600 m, *Parris 10814* (K); PARK HEADQUARTERS: 1500 m, *Parris 11426* (K).

16.4.3. Gonocormus minutus (Blume) Bosch, Hymen. Jav., 7, t. 3 (1861).

Trichomanes minutum Blume, Enum. Pl. Javae, 223 (1828). C. Chr. & Holttum, Gardens' Bull. 7: 216 (1934). *Trichomanes proliferum* Blume, Enum. Pl. Javae, 224 (1828). C. Chr. & Holttum, Gardens' Bull. 7: 216 (1934).

Epiphyte. Hill dipterocarp forest to upper montane forest. Elevation: 600–3400 m.

Material examined: GURULAU SPUR: 1500 m, *Gibbs 401* (K); LIWAGU RIVER TRAIL: 1400 m, *Parris & Croxall 8621* (K); MEKEDEU VALLEY/KING GEORGE PEAK: 2700–3200 m, *Kokawa & Hotta 3526* (K); MEMPENING TRAIL: 1600 m, *Parris 11497* (K); PARK HEADQUARTERS: 1500 m, *Parris 10782* (K), 1500 m, *11421* (K); PORING HOT SPRINGS: 600–900 m, *Kokawa & Hotta 4837* (L); PORING HOT SPRINGS/LANGANAN WATER FALLS: 1000 m, *Parris & Croxall 8932* (K); SUMMIT TRAIL: 2300 m, *Parris & Croxall 8674* (K), 3000 m, *8732* (K), 3300 m, *8768* (K), 3400 m, *8777* (K), 3000 m, *8792* (K); TENOMPOK/LUMU-LUMU: 1500 m, *Holttum SFN 25434* (K, SING).

16.4.4. Gonocormus novoguineensis (Brause) Copel., Philipp. J. Sci. 73: 467 (1941).

Epiphyte. Lower montane forest. Elevation: 1400–1800 m.

Material examined: LIWAGU RIVER TRAIL: 1400 m, *Parris & Croxall 9159* (K); MEMPENING TRAIL: 1600 m, *Parris 11512* (K); SUMMIT TRAIL: 1800 m, *Parris & Croxall 8880* (K).

16.4.5. Gonocormus teijsmanni Bosch, Hymen. Jav., 10, t. 5 (1861).

Trichomanes teysmannii (Bosch) Bosch, Ned. Kr. Arch. 5(2): 142 (1861). C. Chr. & Holttum, Gardens' Bull. 7: 216 (1934).

Epiphyte? Upper montane forest. Elevation: 3000 m.

Material examined: PAKA-PAKA CAVE: 3000 m, *Holttum SFN 25498* (K, SING).

16.4.6. Gonocormus sp. 1

Epiphyte and rupestral. Lower montane forest. Elevation: 1400–2600 m. Endemic to Mount Kinabalu.

Material examined: LAYANG-LAYANG: 2600 m, *Parris 11671* (K); LIWAGU RIVER TRAIL: 1400 m, *Parris & Croxall 8616* (K), 1400 m, *8617* (K).

16.5. HYMENOPHYLLUM

Copeland, E. B. 1937. *Hymenophyllum.* Philipp. J. Sci. 64: 77–93. Copeland, E. B. 1938. Genera Hymenophyllacearum. Philipp. J. Sci. 67: 37–39.

16.5.1. Hymenophyllum peltatum (Poir.) Desv., Prodr., 333 (1827).

Hymenophyllum perfissum Copel., Philipp. J. Sci. 12C: 47 (1917). C. Chr. & Holttum, Gardens' Bull. 7: 214 (1934). Type: Paka-paka Cave: *Clemens 10588* (holotype PNH?†; isotype MICH!).

Lithophyte and epiphyte. Upper montane forest. Elevation: 2900–3300 m.

Additional material examined: JANET'S HALT/SHEILA'S PLATEAU: 2900 m, *Collenette 21542* (L, US); SUMMIT TRAIL: 3000 m, *Parris & Croxall 8734* (K), 3300 m, *8748* (K), 3200 m, *8756* (K).

16.6. MACROGLENA

Copeland, E. B. 1933. *Trichomanes*. Philipp. J. Sci. 51: 258–270. Copeland, E. B. 1938. Genera Hymenophyllacearum. Philipp. J. Sci. 67: 82–85.

16.6.1. Macroglena meifolia (Bory ex Willd.) Copel., Philipp. J. Sci. 67: 83 (1938).

Trichomanes pluma Hook., Ic. Pl., t. 997. C. Chr. & Holttum, Gardens' Bull. 7: 218 (1934).

Terrestrial. Lower montane forest. Elevation: 1500–2400 m.

Material examined: EASTERN SHOULDER: 2000 m, *RSNB 175* (K, SING); KEMBURONGOH: 2400 m, *Clemens 28985* (K); KEMBURONGOH/LUMU-LUMU: 1800 m, *Holttum SFN 25469* (K, SING); LUMU-LUMU: 1800 m, *Clemens 27071* (K); MARAI PARAI: *Topping 1879* (SING); MARAI PARAI SPUR: *Clemens 11056* (K); MENTEKI RIVER: 1600 m, *Beaman 10777* (MICH); MOUNT KINABALU: 1500–1800 m, *Burbidge s.n.* (K), *Low s.n.* (K); MT. LENAU (BETWEEN TENOMPOK AND KEMBURONGOH): 1500 m, *Sinclair et al. 9018* (K, SING); PARK HEADQUARTERS: 1500 m, *Parris & Croxall 8575* (K).

16.6.2. Macroglena obtusa Copel., Philipp. J. Sci. 84: 163 (1955).

Terrestrial. Lower montane oak forest. Elevation: 1500 m.

Material examined: PARK HEADQUARTERS: 1500 m, *Parris & Croxall 8577* (K).

16.6.3. Macroglena schlechteri (Brause) Copel., Philipp. J. Sci. 67: 84 (1938).

Trichomanes schlechteri Brause, Engler's Bot. Jahrb. 49: 10 (1912). C. Chr. & Holttum, Gardens' Bull. 7: 218 (1934).

Elevation: 1800 m.

Material examined: MOUNT KINABALU: 1800 m, *Burbidge s.n.* (K).

16.7. MECODIUM

Copeland, E. B. 1937. *Hymenophyllum*. Philipp. J. Sci. 64: 93–163. Copeland, E. B. 1938. Genera Hymenophyllacearum. Philipp. J. Sci. 67: 17–27.

16.7.1. Mecodium badium (Hook. & Grev.) Copel., Philipp. J. Sci. 67: 23 (1938).

Hymenophyllum badium Hook. & Grev., Ic. Fil., t. 76 (1828). C. Chr. & Holttum, Gardens' Bull. 7: 214 (1934).

Epiphyte. Hill forest. Elevation: 800 m.

Material examined: DALLAS: 800 m, *Clemens 27315* (K).

16.7.2. Mecodium emarginatum (Sw.) Copel., Philipp. J. Sci. 67: 20 (1938).

Hymenophyllum eximium Kunze, Bot. Zeit. 4: 478 (1846). C. Chr. & Holttum, Gardens' Bull. 7: 213 (1934).

Epiphyte. Lower montane to upper montane forest. Elevation: 1400–3200 m.

Material examined: EASTERN SHOULDER: 2900 m, *RSNB 803* (K, SING), 3000 m, *882* (K); EASTERN SHOULDER, CAMP 4: 2700 m, *RSNB 1150* (K, SING); GOLF COURSE SITE: 1700–2000 m, *Beaman 9609* (MICH); GURULAU SPUR: 3200 m, *Clemens 50894* (US); KEMBURONGOH: 2700 m, *Clemens 27925* (K), 2400 m, *28010* (US); KIAU VIEW TRAIL: 1600 m, *Parris 10812* (K); KINATEKI RIVER: 2400 m, *Clemens 31783* (K); LAYANG-LAYANG: 2400–2600 m, *Hotta 3825* (K); LIWAGU RIVER TRAIL: 1400 m, *Parris & Croxall 8615* (K); LUBANG: *Topping 1771* (SING, US); LUBANG/KEMBURONGOH: *Topping 1641* (SING, US); LUMU-LUMU: 1700 m, *Clemens 28388* (K), 2100 m, *29715* (K); MARAI PARAI: 2100–2700 m, *Clemens 33218* (K); MESILAU CAVE TRAIL: 1600–2000 m, *Kokawa & Hotta 4076* (K); PARK HEADQUARTERS: 1500 m, *Parris & Croxall 8579* (K); PINOSUK PLATEAU: 3000 m, *RSNB 882* (SING); SUMMIT TRAIL: 2000 m, *Parris & Croxall 8681* (K), 2600 m, *8822* (K); TENOMPOK: 1500 m, *Clemens 26948* (K), 1500 m, *27791* (US), 1500 m, *27991* (K), 1600 m, *29488* (US), 1400 m, *Holttum SFN 25424* (K, SING); ULAR HILL TRAIL: 1700 m, *Parris 11450* (K), 1700 m, *11456* (K).

16.7.3. Mecodium javanicum (Spreng.) Copel., Philipp. J. Sci. 67: 20 (1938).

Hymenophyllum australe sensu C. Chr. & Holttum non Willd., Gardens' Bull. 7: 213 (1934).

Epiphyte. Lower montane to upper montane forest. Elevation: 1400–3200 m.

Material examined: EASTERN SHOULDER: 3200 m, *RSNB 773* (K, SING); GOKING'S VALLEY: 2700 m, *Fuchs 21464* (K); GURULAU SPUR: 3200 m, *Clemens 50893* (K); KEMBURONGOH/LAYANG-LAYANG: 2400 m, *Edwards 2201* (K); KILEMBUN RIVER: *Clemens 33893* (L, SING); LAYANG-LAYANG/PAKA CAVE: 1800–2600 m, *Hotta 3789* (US); MARAI PARAI: 1400 m, *Clemens 32383* (L); MESILAU/BAMBANGAN RIVERS: 1600–1700 m, *Kokawa & Hotta 4273* (US); NUMERUK CREEK: 1400 m, *Clemens 40051* (K); PAKA-PAKA CAVE: 3000 m, *Holttum SFN 25501* (K, SING); SUMMIT TRAIL: 2500 m, *Parris 11607* (K), 2300 m, *Parris & Croxall 8672* (K), 2200 m, *8680* (K), 2300 m, *8694* (K), 2500 m, *9163* (K); TENOMPOK: 1400 m, *Holttum SFN 25734* (SING).

16.7.4. Mecodium paniculiflorum (C. Presl) Copel., Philipp. J. Sci. 67: 19 (1938).

Hymenophyllum paniculiflorum C. Presl, Hymen., 32, 55 (1843). C. Chr. & Holttum, Gardens' Bull. 7: 213 (1934).

Epiphyte. Upper montane forest, probably mostly on ultramafics. Elevation: 2100–3800 m.

Material examined: GURULAU SPUR: 3700 m, *Clemens 51543* (K, MICH); LUMU-LUMU: 2100 m, *Clemens 28011* (K, SING); PAKA-PAKA CAVE: *Holttum SFN 25497* (K, SING); SUMMIT TRAIL: 3100 m, *Parris 11568* (K), 3000 m, *Parris & Croxall 8705* (K), 3100 m, *8706* (K), 3000 m, *8736* (K), 3000 m, *9161* (K).

16.7.5. Mecodium polyanthos (Sw.) Copel., Philipp. J. Sci. 67: 19 (1938).

Hymenophyllum blumeanum Spreng., Syst. 4: 131 (1827). C. Chr. & Holttum, Gardens' Bull. 7: 213 (1934).

Epiphyte. Hill dipterocarp forest. Elevation: 1000–1500 m.

Material examined: MAMUT RIVER: 1500 m, *RSNB 1743* (SING); PORING HOT SPRINGS/LANGANAN WATER FALLS: 1000 m, *Parris & Croxall 8975* (K).

16.7.6. Mecodium productum (Kunze) Copel., Philipp. J. Sci. 67: 20 (1938).

Hymenophyllum productum Kunze, Bot. Zeit. 6: 305 (1848). C. Chr. & Holttum, Gardens' Bull. 7: 214 (1934).

Epiphyte. Lower montane forest. Elevation: 1100–2300 m.

Material examined: KILEMBUN BASIN: 2300 m, *Clemens 33718* (K, SING); LIWAGU RIVER TRAIL: 1400 m, *Parris & Croxall 9042* (K); LUMU-LUMU: 1900 m, *Sinclair et al. 9207* (K, SING); MAMUT/BAMBANGAN RIVERS: 1600–1800 m, *Kokawa & Hotta 5822* (K); MESILAU TRAIL: 1400–1600 m, *Kokawa & Hotta 3972* (K); PENATARAN BASIN: 1100 m, *Clemens 34104* (K, SING); PINOSUK PLATEAU: 1400–1600 m, *Kokawa & Hotta 4434* (L); TENOMPOK: 1500 m, *Clemens 28228* (SING); TINEKUK FALLS: 1800 m, *Clemens 40900* (K).

16.7.7. Mecodium subdemissum (Christ) Parris, comb. nov.

Hymenophyllum subdemissum Christ, Bull. Boiss. 6: 140 (1898).

Epiphyte. Lower montane oak forest. Elevation: 1400–1600 m.

Material examined: KIAU VIEW TRAIL: 1600 m, *Parris 10816* (K); LIWAGU RIVER TRAIL: 1400 m, *Parris & Croxall 8622* (K), 1400 m, *8626* (K); PARK HEADQUARTERS: 1400 m, *Edwards 2168B* (K), 1500 m, *Parris 10783* (K), 1500 m, *11423* (K), 1500 m, *Parris & Croxall 9157* (K).

16.7.8. Mecodium thuidium (Harr.) Copel., Philipp. J. Sci. 67: 20 (1938).

Hymenophyllum thuidium Harr., J. Linn. Soc. 16: 25 (1877). C. Chr. & Holttum, Gardens' Bull. 7: 215 (1934)

Epiphyte. Lower montane oak forest. Elevation: 1400–1500 m.

Material examined: LIWAGU RIVER TRAIL: 1400 m, *Parris & Croxall 8624* (K); PINOSUK PLATEAU: 1500 m, *Parris & Croxall 9139* (K); TENOMPOK: 1400 m, *Holttum SFN 25392* (SING).

16.7.9. Mecodium treubii (Racib.) Copel., Philipp. J. Sci. 67: 22 (1938).

Epiphyte. Lower montane forest. Elevation: 1200–1500 m.

Material examined: LIWAGU RIVER TRAIL: 1400 m, *Parris & Croxall 8623* (K); MAMUT RIVER: 1200 m, *RSNB 1739* (K, SING); PARK HEADQUARTERS: 1500 m, *Parris 10829* (K).

16.8. MERINGIUM

Copeland, E. B. 1937. *Hymenophyllum*. Philipp. J. Sci. 64: 14–68. Copeland, E. B. 1938. Genera Hymenophyllacearum. Philipp. J. Sci. 67: 39–46.

16.8.1. Meringium acanthoides (Bosch) Copel., Philipp. J. Sci. 67: 42 (1938).

Hymenophyllum australe Willd. var. *fimbriatum* (J. Sm. ex Hook.) C. Chr. sensu C. Chr. & Holttum, Gardens' Bull. 7: 213 (1934) non *Hymenophyllum fimbratum* J. Sm. ex Hook.

Epiphyte. Lower montane forest. Elevation: 1400–3100 m.

Material examined: KIAU VIEW TRAIL: 1600 m, *Parris 10817* (K); KINASARABAN HILL: 1400 m, *Sinclair et al. 8970* (SING); MARAI PARAI: 2100 m, *Clemens 32018* (SING), 2100 m, *33213* (K); MEMPENING TRAIL: 1600 m, *Parris 11501* (K); PAKA-PAKA CAVE: 3100 m, *Edwards 2211* (K); PARK HEADQUARTERS/POWER STATION: 1700–1900 m, *Kokawa & Hotta 3233* (K); TENOMPOK: 1500 m, *Clemens 27425* (BM); ULAR HILL TRAIL: 1700 m, *Parris 11446* (K), 1700 m, *11447* (K).

16.8.2. Meringium bakeri (Copel.) Copel., Philipp. J. Sci. 67: 40 (1938).

Hymenophyllum bakeri Copel., Sarawak Mus. J. 2: 309 (1917). C. Chr. & Holttum, Gardens' Bull. 7: 214 (1934).

Epiphyte. Hill dipterocarp forest (rarely) to upper montane forest. Elevation: 900–2700 m.

Material examined: KEMBURONGOH: 1800–2400 m, *Clemens 28005* (K); KILEMBUN RIVER: 1400 m, *Clemens 33892* (K); KUNDASANG: 1400 m, *Cox 2530* (K); LIWAGU RIVER TRAIL: 1400 m,

Parris & Croxall 8632 (K); MARAI PARAI: 1500 m, *Clemens 32751* (K, SING); PARK HEADQUARTERS: 1500 m, *Parris & Croxall 8581* (K), 1500 m, *8582* (K), 1500 m, *8588* (K); PORING HOT SPRINGS/LANGANAN WATER FALLS: 900 m, *Parris & Croxall 8977* (K); SUMMIT TRAIL: 2200–2500 m, *Parris & Croxall 8676* (K), 2200–2500 m, *8677* (K), 2100 m, *8688* (K), 2200 m, *8693* (K), 2700 m, *8731* (K), 2500 m, *8799* (K), 2700 m, *8838* (K); TENOMPOK: 1700 m, *Clemens 28088* (K, SING).

16.8.3. Meringium blandum (Racib.) Copel., Philipp. J. Sci. 67: 43 (1938).

Hymenophyllum blandum Racib., Pterid. Buit., 20 (1898). C. Chr. & Holttum, Gardens' Bull. 7: 214 (1934).

Epiphyte. Lower montane forest. Elevation: 1500–2700 m.

Material examined: KEMBURONGOH: 2400–2700 m, *Gibbs 4228* (K); KIAU VIEW TRAIL: 1600 m, *Parris 10819* (K); KINASARABAN HILL: 1500 m, *Sinclair et al. 8978* (K, SING); MARAI PARAI: 1500 m, *Clemens 32716* (K), 1500 m, *Holttum SFN 25605* (SING); MARAI PARAI SPUR: 1500 m, *Holttum SFN 25605* (K); MEMPENING TRAIL: 1600 m, *Parris 11496* (K); PARK HEADQUARTERS: 1500 m, *Parris & Croxall 8761* (K); SUMMIT TRAIL: 2000 m, *Parris & Croxall 8687* (K).

16.8.4. Meringium cardunculus (C. Chr.) Copel., Philipp. J. Sci. 67: 42 (1938).

Hymenophyllum cardunculus C. Chr. in Irmsch., Mitt. Inst. Allg. Bot. Hamburg 7: 144 (1928). C. Chr. & Holttum, Gardens' Bull. 7: 215 (1934).

Epiphyte. Lower montane forest. Elevation: 1100–2700 m.

Material examined: DALLAS/TENOMPOK: 1200 m, *Holttum SFN 25351* (SING); GURULAU SPUR: 1500 m, *Gibbs 4020* (K); KIAU VIEW TRAIL: 1600 m, *Parris 10820* (K); KILEMBUN BASIN: 1100 m, *Clemens 34477* (K, SING); LIWAGU RIVER TRAIL: 1400 m, *Parris & Croxall 8627* (K), 1400 m, *9160* (K); PARK HEADQUARTERS: 1500 m, *Parris & Croxall 8586* (K), 1500 m, *8587* (K); PINOSUK PLATEAU: 1500 m, *Parris & Croxall 9140* (K); SUMMIT TRAIL: 2000 m, *Parris & Croxall 8692* (K), 2500 m, *8795* (K), 2700 m, *8839* (K); TENOMPOK: 1200 m, *Holttum SFN 25351* (K).

16.8.5. Meringium denticulatum (Sw.) Copel., Philipp. J. Sci. 67: 42 (1938).

Hymenophyllum denticulatum Sw., Schrad. J. Bot. 1800 (2): 100 (1801). C. Chr. & Holttum, Gardens' Bull. 7: 215 (1934). *Hymenophyllum neesii* (Blume) Hook., Sp. Fil. 1: 99 (1844). C. Chr. & Holttum, Gardens' Bull. 7: 215 (1934). *Hymenophyllum serrulatum* sensu C. Chr. & Holttum p.p. non (C. Presl) C. Chr., Gardens' Bull. 7: 215 (1934).

Epiphyte. Hill dipterocarp forest to lower montane forest, rarely apparently to upper montane forest. Elevation: 600–3000 m.

Material examined: BUNDU TUHAN: 600 m, *Gibbs 3962* (K); EASTERN SHOULDER: 2700 m, *RSNB 948* (K, SING); GOKING'S VALLEY: 2700 m, *Fuchs 21463* (L); GURULAU SPUR: 1500 m, *Clemens 50681* (K), 2400 m, *51113* (K); KINATEKI RIVER: 2700 m, *Clemens 31804* (K, SING); LIWAGU RIVER TRAIL: 1400 m, *Parris & Croxall 8630* (K); LUBANG/PAKA-PAKA CAVE: 2100–3000 m, *Gibbs 4144* (K), *Topping 1721* (US); LUMU-LUMU: 1800 m, *Clemens 28386* (K); MAMUT RIVER: 1200 m, *RSNB 1666* (K, SING), 1200 m, *1736* (K, SING); MARAI PARAI: *Topping 1881* (SING); MARAI PARAI SPUR: *Topping 1881* (US); MESILAU BASIN: 1500–2700 m, *Clemens 28397* (US); MOUNT KINABALU: *Burbidge s.n.* (K), *Low s.n.* (K); PARK HEADQUARTERS/POWER STATION: 1700–1900 m, *Kokawa & Hotta 3229* (L); PORING HOT SPRINGS/LANGANAN WATER FALLS: 900 m, *Parris & Croxall 8978* (K); TENOMPOK: 1500 m, *Clemens 28084* (K); TENOMPOK/LUMU-LUMU: 1500 m, *Holttum SFN 25425* (K, SING).

16.8.6. Meringium holochilum (Bosch) Copel., Philipp. J. Sci. 67: 41 (1938).

Hymenophyllum holochilum (Bosch) C. Chr., Index Fil., 362 (1905). C. Chr. & Holttum, Gardens' Bull. 7: 214 (1934). *Hymenophyllum serrulatum* (C. Presl) C. Chr., Index Fil., 367 (1905). C. Chr. & Holttum, Gardens' Bull. 7: 215 (1934) p.p.

Epiphyte. Lower montane forest, doubtfully as high as the *Clemens 51397* record. Elevation: 1100–3700 m.

Material examined: EASTERN SHOULDER: 1100 m, *RSNB 673* (K, SING), 2700 m, *1142* (SING), 1200 m, *1594* (SING); EASTERN SHOULDER, CAMP 4: 2700 m, *RSNB 1142* (K); GURULAU SPUR: 1700 m, *Clemens 50697* (K); KIAU VIEW TRAIL: 1600 m, *Parris 10815* (K), 1600 m, *11613* (K), 1600 m, *Parris & Croxall 8558* (K); KILEMBUN BASIN: 1400 m, *Clemens s.n.* (K); KINASARABAN HILL: 1500 m, *Sinclair et al. 8981* (K); KINATEKI RIVER HEAD: 2700 m, *Clemens 32250* (US); LIWAGU RIVER TRAIL: 1400 m, *Parris & Croxall 8631* (K), 1400 m, *8633* (K); MARAI PARAI: 1500 m, *Clemens 32250* (US), 1200 m, *32395* (K, SING), 1500 m, *33162* (K, SING); MEMPENING TRAIL: 1600 m, *Parris 11502* (K); PARK HEADQUARTERS: 1500–1600 m, *Kokawa & Hotta 2942* (K); PENIBUKAN: 1100 m, *Kanis SAN 53975* (K); SUMMIT TRAIL: 2600 m, *Parris 11556* (K); TENOMPOK: 1500 m, *Clemens 26850* (K); VICTORIA PEAK: 3700 m, *Clemens 51397* (K).

16.8.7. **Meringium hosei** (Copel.) Copel., Philipp. J. Sci. 67: 42 (1938).

Epiphyte. Lower montane oak forest. Elevation: 1400–2300 m.

Material examined: KIAU VIEW TRAIL: 1600 m, *Parris 11672* (K); LIWAGU RIVER TRAIL: 1500 m, *Parris 10841* (K), 1400 m, *Parris & Croxall 8628* (K), 1400 m, *8629* (K); PARK HEADQUARTERS: 1400 m, *Edwards 2168A* (K), 1500 m, *Parris 10785* (K), 1500 m, *Parris & Croxall 8583* (K), 1500 m, *8584* (K); SUMMIT TRAIL: 2200–2300 m, *Parris & Croxall 8695* (K).

16.8.8. **Meringium johorense** (Holttum) Copel., Fern Fl. Philipp. 1: 64 (1958).

Hymenophyllum johorense Holttum, Gardens' Bull. 4: 408 (1929). C. Chr. & Holttum, Gardens' Bull. 7: 214 (1934).

Epiphyte. Upper montane forest. Elevation: 2700 m.

Material examined: KEMBURONGOH/PAKA-PAKA CAVE: 2700 m, *Holttum SFN 25482* (K); SUMMIT TRAIL: 2700 m, *Parris & Croxall 8840* (K).

16.8.9. **Meringium lobbii** (T. Moore ex Bosch) Copel., Philipp. J. Sci. 67: 43 (1938).

Hymenophyllum lobbii T. Moore ex Bosch, Ned. Kr. Arch. 5, 3: 176 (1863). C. Chr. & Holttum, Gardens' Bull. 7: 215 (1934).

Epiphyte. Hill dipterocarp forest to lower montane forest. Elevation: 900–2400 m.

Material examined: KEMBURONGOH: 2400 m, *Clemens 28006* (K, SING); PORING HOT SPRINGS/LANGANAN WATER FALLS: 900 m, *Parris & Croxall 8979* (K); SUMMIT TRAIL: 2100 m, *Parris 11611* (K), 2000 m, *Parris & Croxall 8686* (K).

16.8.10. **Meringium macrosorum** (Alderw.) Copel., Philipp. J. Sci. 67: 43 (1938).

Epiphyte. Upper montane forest. Elevation: 3000–3400 m.

Material examined: PAKA-PAKA CAVE: 3100 m, *Parris 11580* (K); PANAR LABAN: 3300 m, *Parris 11586* (K), 3400 m, *Parris & Croxall 8776* (K); SUMMIT TRAIL: 3100 m, *Parris & Croxall 8707* (K), 3000 m, *8708* (K), 3000 m, *8711* (K), 3000 m, *8712* (K), 3000 m, *8733* (K), 3000 m, *8735* (K), 3300 m, *8749* (K), 3200 m, *8775* (K).

16.8.11. **Meringium microchilum** (Baker) Parris, comb. nov.

Trichomanes microchilum Baker, Trans. Linn. Soc. Bot. 4: 250 (1894). Type: Mount Kinabalu: 2100 m, *Haviland 1478* (holotype K!). *Hymenophyllum microchilum* (Baker) C. Chr. in Irmsch., Mitt. Inst. Allg. Bot. Hamburg 7: 143 (1928). C. Chr. & Holttum, Gardens' Bull. 7: 212 (1934).

Epiphyte. Lower montane forest. Elevation: 1200–2400 m.

Additional material examined: KEMBURONGOH: 2400 m, *Clemens 28984* (MICH, SING); KEMBURONGOH/LUMU-LUMU: 1800–2100 m, *Holttum SFN 25470* (K, SING); KIAU VIEW TRAIL: 1600 m, *Parris 10821* (K); LIWAGU RIVER TRAIL: 1800 m, *Parris 11465* (K), 1400 m, *Parris &*

Croxall 8625 (K); LUMU-LUMU: 1800 m, *Clemens 27070* (K, MICH); PARK HEADQUARTERS: 1500 m, *Parris 10784* (K), 1500 m, *11436* (K), 1500 m, *Parris & Croxall 8559* (K), 1500 m, *8585* (K), 1500 m, *9025* (K); PENIBUKAN: 1200–1500 m, *Clemens 31145* (K); PINOSUK PLATEAU: 1500 m, *Parris & Croxall 9141* (K); SUMMIT TRAIL: 2000 m, *Parris & Croxall 8682* (K); TENOMPOK: 1500 m, *Holttum SFN 25735* (SING).

16.8.12. Meringium pachydermicum (Ces.) Copel., Philipp. J. Sci. 67: 41 (1938).

Hymenophyllum pachydermicum Ces., Atti Ac. Napoli 7, 8: 8 (1876). C. Chr. & Holttum, Gardens' Bull. 7: 213 (1934). *Hymenophyllum clemensiae* Copel., Philipp. J. Sci. 12C: 46 (1917). Type: Gurulau Spur: *Clemens 10780* (holotype PNH?†; isotype MICH!).

Epiphyte. Lower montane forest. Elevation: 900–1700 m.

Additional material examined: GURULAU SPUR: *Topping 1619* (US); KADAMAIAN RIVER: 900 m, *Holttum SFN 25580* (SING); KADAMAIAN RIVER NEAR MINITINDUK: 900 m, *Holttum SFN 25580* (K); KIAU: *Clemens 10226* (MICH); KIAU VIEW TRAIL: 1600 m, *Parris 10818* (K), 1600 m, *Parris & Croxall 8557* (K); KILEMBUN BASIN: 1400 m, *Clemens 34196* (K, SING); LIWAGU RIVER TRAIL: 1500 m, *Parris 10830* (K); MARAI PARAI SPUR: 1500 m, *Holttum SFN 25618* (SING); TENOMPOK: 1700 m, *Clemens 29270* (K, MICH), 1700 m, *Kokawa 6312* (K); TENOMPOK/LUMU-LUMU: 1500 m, *Holttum SFN 25426* (K); ULAR HILL TRAIL: 1700 m, *Parris 11444* (K).

16.8.13. Meringium penangianum (Matthew & Christ ex Christ) Copel., Philipp. J. Sci. 67: 41 (1938).

Epiphyte. Hill dipterocarp forest to upper montane forest. Elevation: 1000–2900 m.

Material examined: KEMBURONGOH/PAKA-PAKA CAVE: 2700 m, *Molesworth-Allen 3245* (K); PORING HOT SPRINGS/LANGANAN WATER FALLS: 1000 m, *Parris & Croxall 8976* (K); POWER STATION/LAYANG-LAYANG: 2700–2900 m, *Kokawa & Hotta 3731* (K).

16.9. MICROGONIUM

Croxall, J. P. 1986. *Microgonium* (Hymenophyllaceae) in Malesia, with special reference to Peninsular Malaysia. Kew Bull. 41: 519–531.

16.9.1. Microgonium bimarginatum Bosch, Hymen. Jav., 7 (1861).

Lithophyte. Lowland dipterocarp forest. Elevation: 600 m.

Material examined: BAT CAVE: 600 m, *Parris & Croxall 8969* (K).

16.9.2. Microgonium sublimbatum (C. Muell.) Bosch, Hymen. Jav., 6, pl. 2 (1861).

Epiphyte. Lowland dipterocarp forest. Elevation: 600 m.

Material examined: KIPUNGIT FALLS: 600 m, *Parris & Croxall 8918* (K).

16.10. MICROTRICHOMANES

16.10.1. Microtrichomanes dichotomum (Kunze) Copel., Philipp. J. Sci. 67: 36 (1938).

Trichomanes dichotomum Kunze, Bot. Zeit. 6: 285 (1848). C. Chr. & Holttum, Gardens' Bull. 7: 216 (1934) p.p.

Epiphyte. Lower montane forest. Elevation: 2300–2700 m.

Material examined: GURULAU SPUR: 2400 m, *Clemens 51115* (K); KEMBURONGOH/PAKA-PAKA CAVE: 2400–2700 m, *Holttum SFN 25475* (K); KINATEKI RIVER HEAD: 2400–2700 m, *Clemens 31960* (K); LAYANG-LAYANG: 2600 m, *Parris 11600* (K); SUMMIT TRAIL: 2300 m, *Parris 11537* (K), 2300 m, *Parris & Croxall 8673* (K), 2500 m, *8724* (K).

16.10.2. Microtrichomanes digitatum (Sw.) Copel., Philipp. J. Sci. 67: 36 (1938).

Epiphyte. Lower montane forest. Elevation: 1200–2700 m.

Material examined: EASTERN SHOULDER: 1200 m, *RSNB 1593* (K, SING); LAYANG-LAYANG: 2400–2600 m, *Hotta 3850* (K); SUMMIT TRAIL: 2700 m, *Parris & Croxall 8841* (K).

16.10.3. Microtrichomanes nitidulum (Blume) Copel., Philipp. J. Sci. 67: 37 (1938).

Trichomanes dichotomum sensu C. Chr. & Holttum p.p. non Kunze, Gardens' Bull. 7: 216 (1934).

Epiphyte. Lower and upper montane forest. Elevation: 1500–3100 m.

Material examined: EASTERN SHOULDER: 3100 m, *RSNB 850* (K, SING); LIWAGU RIVER TRAIL: 1800 m, *Parris 11468* (K); LUBANG: 2100 m, *Gibbs 4253* (K); MEMPENING TRAIL: 1600 m, *Parris 11511* (K); PARK HEADQUARTERS: 1500 m, *Parris & Croxall 8578* (K); SUMMIT TRAIL: 1900 m, *Parris & Croxall 8683* (K), 2100 m, *8684* (K), 3100 m, *8713* (K).

16.10.4. Microtrichomanes palmatifidum (C. Muell.) Copel., Philipp. J. Sci. 67: 36 (1938).

Trichomanes palmatifidum C. Muell., Bot. Zeit. 12: 732 (1854). C. Chr. & Holttum, Gardens' Bull. 7: 216 (1934).

Epiphyte. Lower and upper montane forest. Elevation: 1800–3000 m.

Material examined: DACHANG: 3000 m, *Clemens 28207* (K); EASTERN SHOULDER: 2000 m, *RSNB 2005* (K, SING); GURULAU SPUR: 3000 m, *Clemens 50890* (K); KEMBURONGOH/LUMU-LUMU: 1800 m, *Holttum SFN 25462* (K, SING); KILEMBUN RIVER: 2700–3000 m, *Clemens 33927* (K); KILEMBUN RIVER HEAD: 2700–3000 m, *Clemens 33927* (SING); SUMMIT TRAIL: 2200 m, *Parris & Croxall 8685* (K).

16.10.5. Microtrichomanes ridleyi (Copel.) Copel., Philipp. J. Sci. 67: 36 (1938).

Epiphyte. Lower montane mossy forest. Elevation: 2400–2600 m.

Material examined: KEMBURONGOH/LAYANG-LAYANG: 2400 m, *Edwards 2222* (K); LAYANG-LAYANG: 2600 m, *Parris 11602* (K); SUMMIT TRAIL: 2500 m, *Parris & Croxall 8670* (K), 2500 m, *9176* (K).

16.10.6. Microtrichomanes vitiense (Baker) Copel., Philipp. J. Sci. 67: 37 (1938).

Epiphyte. Lowland dipterocarp forest. Elevation: 600 m.

Material examined: PORING HOT SPRINGS/LANGANAN WATER FALLS: 600 m, *Parris & Croxall 8974* (K).

16.11. NESOPTERIS

Copeland, E. B. 1933. *Trichomanes*. Philipp. J. Sci. 51: 223–227. Copeland, E. B. 1938. Genera Hymenophyllacearum. Philipp. J. Sci. 67: 65–66.

16.11.1. Nesopteris grandis (Copel.) Copel., Philipp. J. Sci. 67: 66 (1938).

Terrestrial. Lowland dipterocarp forest. Elevation: 700 m.

Material examined: PORING HOT SPRINGS/LANGANAN WATER FALLS: 700 m, *Parris &
Croxall 8973* (K).

16.12. PLEUROMANES

Copeland, E. B. 1933. *Trichomanes*. Philipp. J. Sci. 51: 138–143. Copeland, E. B. 1938. Genera
Hymenophyllacearum. Philipp. J. Sci. 67: 55–57.

16.12.1. Pleuromanes album (Blume) Parris, comb. nov.

Trichomanes album Blume, Enum. Pl. Javae, 226 (1828). C. Chr. & Holttum,
Gardens' Bull. 7: 217 (1934).

Epiphyte. Lower montane oak forest. Elevation: 1500–2500 m.

Material examined: KEMBURONGOH: 2300 m, *Clemens 28007* (K); KEMBURONGOH/LUMU-
LUMU: 1800 m, *Holttum SFN 25468* (K); LUBANG/KEMBURONGOH: 1800–2100 m, *Gibbs 4143* (K);
PARK HEADQUARTERS: 1500 m, *Parris & Croxall 9164* (K); SUMMIT TRAIL: 2500 m, *Parris 11606*
(K).

16.12.2. Pleuromanes pallidum (Blume) C. Presl, Epim., 258 (1849).

Trichomanes pallidum Blume, Enum. Pl. Javae, 225 (1828). C. Chr. & Holttum,
Gardens' Bull. 7: 217 (1934).

Epiphyte. Lower montane forest. Elevation: 1200–2800 m.

Material examined: EASTERN SHOULDER: 2700 m, *RSNB 907* (K, SING); EASTERN
SHOULDER, CAMP 4: 2700 m, *RSNB 1141* (K, SING); KEMBURONGOH: 2000 m, *Meijer SAN 29215*
(K), 2300 m, *Sinclair et al. 9070* (K, SING); KEMBURONGOH/LUMU-LUMU: 1800 m, *Holttum SFN
25468* (SING); KIBAMBANG RIVER: 1200 m, *Clemens 34372* (K, SING); KILEMBUN BASIN: 1400 m,
Clemens 34195 (K); KINATEKI RIVER: 1200–1500 m, *Clemens 31415* (K); LAYANG-LAYANG/
MESILAU CAVE ROUTE: 2000 m, *Collenette 899* (K); LIWAGU RIVER TRAIL: 1700 m, *Parris 11474*
(K), 1500 m, *Parris & Croxall 8619* (K); LUMU-LUMU: 1500 m, *Holttum SFN 25733* (SING); MAMUT
HILL: 1400–1700 m, *Kokawa & Hotta 5267* (K); MAMUT/BAMBANGAN RIVERS: 1400–1700 m,
Kokawa & Hotta 5557 (L); MARAI PARAI: 1800–2100 m, *Clemens 32951* (K); PARK HEADQUARTERS:
1500 m, *Parris & Croxall 8580* (K); SUMMIT TRAIL: 2800 m, *Collenette 540* (K), 2300 m, *Parris 11538*
(K), 2200 m, *Parris & Croxall 8679* (K), 2500 m, *8720* (K), 2300 m, *8849* (K), 2200 m, *8850* (K), 2500 m,
9162 (K).

16.13. REEDIELLA

Copeland, E. B. 1933. *Trichomanes*. Philipp. J. Sci. 51: 163–174. Copeland, E. B. 1938. Genera
Hymenophyllacearum. Philipp. J. Sci. 67: 57–58.

16.13.1. Reediella endlicheriana (C. Presl) Pichi Serm., Webbia 24: 719 (1970).

Lithophyte. Lower montane forest. Elevation: 1400 m.

Material examined: LIWAGU RIVER TRAIL: 1400 m, *Parris & Croxall 8618* (K).

16.13.2. Reediella humilis (G. Forster) Pichi Serm., Webbia 24: 719 (1970).

Trichomanes humile G. Forster, Prodr., 84 (1786). C. Chr. & Holttum, Gardens'
Bull. 7: 216 (1934).

Epiphyte. Hill forest. Elevation: 800 m.

Material examined: DALLAS: 800 m, *Clemens 27138* (K); MAYAMUT RIVER: *Holttum SFN 25375*
(K, SING).

16.14. SELENODESMIUM

Copeland, E. B. 1933. *Trichomanes*. Philipp. J. Sci. 51: 232–245. Copeland, E. B. 1938. Genera Hymenophyllacearum. Philipp. J. Sci. 67: 80–81.

16.14.1. Selenodesmium obscurum (Blume) Copel., Philipp. J. Sci. 67: 81 (1938).

Trichomanes cupressoides sensu C. Chr. & Holttum non Desv., Gardens' Bull. 7: 218 (1934). *Trichomanes papillatum* C. Muell., Bot. Zeit. 12: 751 (1854). C. Chr. & Holttum, Gardens' Bull. 7: 217 (1934).

Terrestrial. Hill forest (rarely) to lower montane forest. Elevation: 500–2700 m.

Material examined: EASTERN SHOULDER: 2700 m, *RSNB 951* (K, SING); KIAU VIEW TRAIL: 1600 m, *Parris 10813* (K); KILEMBUN RIVER: 1400 m, *Clemens 32513* (K), 1400 m, *40014* (K); LIWAGU RIVER TRAIL: 1700 m, *Parris 11471* (K), 1400 m, *Parris & Croxall 8614* (K); MAMUT/BAMBANGAN RIVERS: 1400–1700 m, *Kokawa & Hotta 5556* (K); MARAI PARAI SPUR: *Topping 1847* (SING), *1877* (SING); NUMERUK CREEK: 1500 m, *Clemens 40040* (K); PARK HEADQUARTERS: 1400 m, *Edwards 2164* (K), 1500–1600 m, *Kokawa & Hotta 2944* (K), 1500 m, *Parris 11435* (K), 1500 m, *Parris & Croxall 8576* (K); PENIBUKAN: 1200 m, *Clemens 30741* (K), 1500 m, *31175* (K), 1200–1500 m, *31251* (K), 1200–1500 m, *Kanis SAN 53988* (K), 1300 m, *Parris 11524* (K); PINOSUK PLATEAU: 1600 m, *Parris & Croxall 9150* (K); PORING HOT SPRINGS: 800 m, *Meijer SAN 24039* (K), 500 m, *Parris & Croxall 8992* (K); SUMMIT TRAIL: 2300 m, *Parris & Croxall 8671* (K); TENOMPOK: 1500 m, *Clemens 28225* (K), 1700 m, *28459* (K), 1400 m, *Holttum SFN 25401* (K, SING); TENOMPOK/LUMU-LUMU: 1500 m, *Holttum SFN 25430* (K, SING).

16.15. SPHAEROCIONIUM

Copeland, E. B. 1937. *Hymenophyllum*. Philipp. J. Sci. 64: 164–176. Copeland, E. B. 1938. Genera Hymenophyllacearum. Philipp. J. Sci. 67: 28–34.

16.15.1. Sphaerocionium pilosissimum (C. Chr.) Copel., Philipp. J. Sci. 67: 33 (1938).

Hymenophyllum pilosissimum C. Chr. in C. Chr. & Holttum, Gardens' Bull. 7: 213 (1934). Type: Mount Kinabalu: *Burbidge s.n.* (holotype K!).

Epiphyte. Lower montane forest. Elevation: 1200–2700 m. Not endemic to Mount Kinabalu.

Additional material examined: KADAMAIAN RIVER/KEMBURONGOH: 2000 m, *Abbe 10196* (K); KEMBURONGOH: 2400 m, *Clemens 28983* (K); KEMBURONGOH/LUMU-LUMU: 1800–2100 m, *Holttum SFN 25460* (K, SING); KEMBURONGOH/PAKA-PAKA CAVE: 2700 m, *Molesworth-Allen 3241* (K); KIAU VIEW TRAIL: 1600 m, *Parris 11618* (K); LAYANG-LAYANG/MESILAU CAVE ROUTE: 2000 m, *Collenette 900* (K); LIWAGU RIVER TRAIL: 1400 m, *Parris & Croxall 9158* (K); MAMUT HILL: 1400–1700 m, *Kokawa & Hotta 5268* (US); MAMUT/BAMBANGAN RIVERS: 1400–1700 m, *Kokawa & Hotta 5505* (K); MESILAU TRAIL: 1700 m, *Hale 28266* (US); MT. LENAU (BETWEEN TENOMPOK AND KEMBURONGOH): 1600 m, *Sinclair et al. 9014* (K, SING); PARK HEADQUARTERS: 1500–1600 m, *Kokawa & Hotta 2836* (US), 1500 m, *Parris & Croxall 8574* (K); PARK HEADQUARTERS/POWER STATION: 1700–1900 m, *Kokawa & Hotta 3228* (K); SOSOPODON/PARK HEADQUARTERS: 1200–1600 m, *Kokawa & Hotta 5151* (K); TENOMPOK: 1500 m, *Clemens 28983* (SING).

16.16. VANDENBOSCHIA

Copeland, E. B. 1933. *Trichomanes*. Philipp. J. Sci. 51: 212–223. Copeland, E. B. 1938. Genera Hymenophyllacearum. Philipp. J. Sci. 67: 51–55.

16.16.1. Vandenboschia auriculata (Blume) Copel., Philipp. J. Sci. 67: 55 (1938).

Trichomanes auriculatum Blume, Enum. Pl. Javae, 225 (1828). C. Chr. & Holttum, Gardens' Bull. 7: 217 (1934).

Terrestrial and epiphytic. Lowland dipterocarp forest to lower montane forest. Elevation: 600–1500 m.

Material examined: DALLAS/TENOMPOK: 1100 m, *Clemens 27736* (K); EASTERN SHOULDER: 900 m, *RSNB 1597* (K, SING); LANGANAN FALLS: 600 m, *Parris & Croxall 8970* (K); TENOMPOK: 1500 m, *Clemens 29347* (K).

16.16.2. Vandenboschia maxima (Blume) Copel., Philipp. J. Sci. 67: 54 (1938).

Trichomanes maximum Blume, Enum. Pl. Javae, 228 (1828). C. Chr. & Holttum, Gardens' Bull. 7: 217 (1934).

Terrestrial. Hill forest to lower montane forest. Elevation: 700–2300 m.

Material examined: DALLAS/TENOMPOK: 1100 m, *Clemens 26897* (K); EASTERN SHOULDER: 700 m, *RSNB 1501* (K, SING); KEMBURONGOH: 2100 m, *Holttum SFN 25541* (K, SING), 2000 m, *Meijer SAN 29214* (K); LIWAGU RIVER TRAIL: 1400 m, *Parris & Croxall 9044* (K); LUBANG/PAKA-PAKA CAVE: *Topping 1752* (SING); MARAI PARAI SPUR: 2100 m, *Gibbs 4036* (K); MESILAU BASIN: 2100 m, *Clemens 29067* (K); MESILAU CAVE TRAIL: 1600–2000 m, *Kokawa & Hotta 4029* (K); MOUNT KINABALU: 1500–1800 m, *Burbidge s.n.* (K), 1800 m, *Low s.n.* (K); POWER STATION/LAYANG-LAYANG: 1700–1900 m, *Kokawa & Hotta 3699* (K); SUMMIT TRAIL: 2300 m, *Parris & Croxall 8669* (K); TENOMPOK: 1500 m, *Clemens 28372* (K).

17. LOMARIOPSIDACEAE

Holttum, R. E. 1978. *Lomariopsis* Group. Fl. Males. 2, 1 (4): 255–330.

17.1. BOLBITIS

Hennipman, E. 1978. *Bolbitis.* Fl. Males. 2, 1 (4): 314–330.

17.1.1. Bolbitis repanda (Blume) Schott, Gen. Fil., ad t. 13 (1835).

Campium quoyanum sensu C. Chr. & Holttum non (Gaud.) Copel., Gardens' Bull. 7: 262 (1934).

Terrestrial. Hill forest. Elevation: 900 m.

Material examined: MINITINDUK: 900 m, *Holttum SFN 25585* (K, SING).

17.1.2. Bolbitis sinuata (C. Presl) Hennipman, Blumea 18: 148 (1970).

Campium subsimplex (Fée) Copel., Philipp. J. Sci. 37: 356, f. 11, pl. 8 (1928). C. Chr. & Holttum, Gardens' Bull. 7: 261 (1934).

Terrestrial. Lower montane forest. Elevation: 1100–1500 m.

Material examined: DALLAS: 1100 m, *Holttum SFN 25141* (SING); DALLAS/TENOMPOK: 1100 m, *Holttum SFN 25352* (K, SING); TENOMPOK: 1500 m, *Clemens 27653* (K).

17.2. ELAPHOGLOSSUM

Holttum, R. E. 1978. *Elaphoglossum.* Fl. Males. 2, 1 (4): 289–314.

17.2.1. Elaphoglossum angulatum (Blume) T. Moore, Index, 5 (1857). C. Chr. & Holttum, Gardens' Bull. 7: 290 (1934).

Epiphyte. Lower and upper montane forest. Elevation: 1900–3800 m.

Material examined: EASTERN SHOULDER: 2700 m, *RSNB 915* (K, SING); GURULAU SPUR: 2100 m, *Clemens 51449* (K); PAKA-PAKA CAVE: 3400 m, *Clemens 28020* (K), 3000 m, *Holttum SFN 25508* (K, SING), 3100 m, *Parris 11578* (K), 3000 m, *Parris & Croxall 8710* (K); SUMMIT AREA: 3800 m, *Collenette 614* (K); SUMMIT TRAIL: 2800 m, *Parris & Croxall 8726* (K), 2500 m, *8816* (K), 1900 m, *8853* (K).

17.2.2. Elaphoglossum annamense C. Chr. & Tardieu-Blot, Not. Syst. Paris 8: 209 (1939).

Elaphoglossum decurrens (Desv.) T. Moore var. *crassum* C. Chr. in C. Chr. & Holttum p.p., Gardens' Bull. 7: 290 (1934). Type: Kemburongoh/Lumu-lumu: 1800 m, *Holttum SFN 25454* (syntype BM n.v.; isosyntypes K!, SING!). *Elaphoglossum beccarianum* sensu C. Chr. & Holttum non (Baker) C. Chr., Gardens' Bull. 7: 289 (1934).

Epiphytic and terrestrial. Lower montane oak forest. Elevation: 1600–2100 m.

Additional material examined: KIAU VIEW TRAIL: 1600 m, *Parris 10799* (K); LIWAGU RIVER TRAIL: 1700 m, *Parris & Croxall 9111* (K); LUMU-LUMU: 1800–2100 m, *Clemens 28019* (K); MESILAU CAVE TRAIL: 1700–1900 m, *Beaman 9112* (MICH); SUMMIT TRAIL: 2100 m, *Parris & Croxall 8689* (K).

17.2.3. Elaphoglossum blumeanum (Fée) J. Sm., Ferns Br. & For., 106 (1866).

Epiphyte. Lower montane oak forest. Elevation: 1400–1800 m.

Material examined: LIWAGU RIVER TRAIL: 1400 m, *Parris & Croxall 8608* (K); MENTEKI RIVER: 1600 m, *Beaman 10787* (MICH, MSC); SUMMIT TRAIL: 1800 m, *Parris & Croxall 8848* (K).

17.2.4. Elaphoglossum callifolium (Blume) T. Moore, Index, 7 (1857). C. Chr. & Holttum, Gardens' Bull. 7: 289 (1934).

Epiphyte. Lower montane forest. Elevation: 1400–2500 m.

Material examined: PINOSUK PLATEAU: 1500 m, *Holttum 29* (SING); POWER STATION: 1800 m, *Holttum 59* (SING); SUMMIT TRAIL: 2500 m, *Parris & Croxall 8805* (K); TENOMPOK: 1400 m, *Holttum SFN 25412* (SING).

17.2.5. Elaphoglossum commutatum (Mett. ex Kuhn) Alderw., Handb. Malayan Ferns Suppl., 427 (1917).

Elaphoglossum callifolium sensu C. Chr. & Holttum p.p. non (Blume) T. Moore, Gardens' Bull. 7: 289 (1934).

Epiphyte. Lower montane forest. Elevation: 1500–1600 m.

Material examined: MOUNT KINABALU: 1500 m, *Shim SAN 75408* (K); PINOSUK PLATEAU: 1600 m, *Parris & Croxall 9152* (K); TENOMPOK: 1500 m, *Clemens 29332* (K).

17.2.6. Elaphoglossum heterolepium Alderw., Bull. Jard. Bot. Buit. 2, 16: 13 (1914).

Elaphoglossum petiolatum (Sw.) Urb. var., C.Chr. & Holttum, Gardens' Bull. 7: 291 (1934).

Epiphyte. Lower montane forest. Elevation: 900–2100 m.

Material examined: BAMBANGAN RIVER: 1500 m, *RSNB 4617* (K); LIWAGU RIVER TRAIL: 1400 m, *Parris & Croxall 9045* (K); LUMU-LUMU: 2100 m, *Clemens 29709* (K); MESILAU BASIN: 1800 m, *Clemens 29064* (K); PARK HEADQUARTERS: 1500 m, *Parris 10833* (K); PINOSUK PLATEAU: 1500 m, *Holttum 28* (SING); SUMMIT TRAIL: 1800 m, *Parris & Croxall 8852* (K); TAHUBANG RIVER: 900 m, *Clemens 40295* (K); TENOMPOK: 1400 m, *Holttum SFN 25357* (K).

17.2.7. Elaphoglossum spongophyllum P. Bell ex Holttum, Blumea 14: 325 (1966). Type: Kinateki River Head: *Clemens 31869* (holotype BO n.v.).

Elaphoglossum decurrens sensu C. Chr. & Holttum p.p. non (Desv.) T. Moore, Gardens' Bull. 7: 290 (1934). *Elaphoglossum decurrens* (Desv.) T. Moore var. *crassum* C. Chr. in C. Chr. & Holttum p.p., Gardens' Bull. 7: 290 (1934). Type: Kemburongoh: 2100 m, *Holttum SFN 25718* (syntype BM n.v.; isosyntype SING!).

Epiphyte. Lower and upper montane forest. Elevation: 2100–2900 m. Not endemic to Mount Kinabalu.

Additional material examined: JANET'S HALT/SHEILA'S PLATEAU: 2900 m, *Collenette 21527* (K); KINATEKI RIVER: 2100–2700 m, *Clemens 31799* (K); SUMMIT TRAIL: 2500 m, *Parris & Croxall 8806* (K).

17.2.8. Elaphoglossum stenolepis P. Bell ex Holttum, Blumea 14: 325 (1966). Type: Lumu-lumu: 1800–2100 m, *Clemens 28019* (holotype US n.v.).

Elaphoglossum decurrens sensu C. Chr. & Holttum p.p. non (Desv.) T. Moore, Gardens' Bull. 7: 290 (1934). *Elaphoglossum decurrens* (Desv.) T. Moore var. *crassum* C. Chr. in C. Chr. & Holttum p.p., Gardens' Bull. 7: 290 (1934). Type: Kemburongoh: 2100 m, *Holttum SFN 25717* (syntype BM n.v.; isosyntypes K!, SING!).

Epiphyte. Lower montane forest. Elevation: 1500–2100 m. Not endemic to Mount Kinabalu.

Additional material examined: KEMBURONGOH: 2000–2100 m, *Holttum 15* (SING); LUMU-LUMU: *Clemens 27060* (K); PARK HEADQUARTERS: 1500 m, *Parris 11623* (K); TENOMPOK/LUMU-LUMU: 1500 m, *Holttum SFN 25427* (SING).

17.3. LOMAGRAMMA

Holttum, R. E. 1978. *Lomagramma*. Fl. Males. 2, 1 (4): 276–289.

17.3.1. Lomagramma brooksii Copel., Philipp.J. Sci. 3C: 345 (1909). C. Chr. & Holttum, Gardens' Bull. 7: 265 (1934).

Terrestrial climber. Hill forest and lower montane forest. Elevation: 800–1700 m.

Material examined: DALLAS: 1100 m, *Holttum SFN 25251* (K, SING, US); KAUNG: 800 m, *Clemens 51308* (K); PENIBUKAN: 1200 m, *Clemens 30636* (K, L, US), 1700 m, *50263* (K, L); TENOMPOK: 1500 m, *Clemens 28638* (K, US).

17.3.2. Lomagramma sinuata C. Chr., Svensk Bot. Tidskr. 16: 98, f. 5 (1922).

Terrestrial climber. Lowland dipterocarp forest. Elevation: 600–900 m.

Material examined: KIPUNGIT FALLS: 600 m, *Parris & Croxall 8904* (K); MINITINDUK: 900 m, *Clemens s.n.* (SING).

17.4. LOMARIOPSIS

Holttum, R. E. 1978. *Lomariopsis*. Fl. Males. 2, 1 (4): 258–263.

17.4.1. Lomariopsis spectabilis (Kunze) Mett., Fil. Hort. Bot. Lips., 22 (1856).

Lomariopsis leptocarpa sensu C. Chr. & Holttum non Fée, Gardens' Bull. 7: 262 (1934).

Terrestrial climber. Hill forest? Elevation: 900–1200 m.

Material examined: DALLAS: 900 m, *Clemens 27596* (K), 900 m, *30457* (K, US); DALLAS/TENOMPOK: 1200 m, *Clemens 27596* (SING, US); EASTERN SHOULDER: 1200 m, *RSNB 38* (K, SING).

17.5. TERATOPHYLLUM

Holttum, R. E. 1978. *Teratophyllum*. Fl. Males. 2, 1 (4): 265–276.

17.5.1. Teratophyllum clemensiae Holttum in C. Chr. & Holttum, Gardens' Bull. 7: 262 (1934). Type: Penibukan: *Clemens 31614* (holotype SING n.v.).

Terrestrial climber. Hill forest and lower montane forest. Elevation: 900–1600 m. Not endemic to Mount Kinabalu.

Additional material examined: DALLAS: 900 m, *Clemens 27144* (K); GURULAU SPUR: 1500 m, *Clemens 50592* (K); KIAU VIEW TRAIL: 1600 m, *Parris & Croxall 9129* (K); LIWAGU RIVER TRAIL: 1500 m, *Parris & Croxall 9046* (K); MT. NUNGKEK: 1200 m, *Clemens 32830* (K); PENIBUKAN: 1200 m, *Clemens 30756* (K), 1200 m, *30809* (SING), 1200 m, *30871* (K), 1200 m, *30890* (K), 1200–1500 m, *31346* (K, SING), 1200–1500 m, *31610* (SING), 1200 m, *31706* (K); TENOMPOK: 1500 m, *Clemens 29566* (SING), 1500 m, *29573* (K).

18. MARATTIACEAE

18.1. ANGIOPTERIS

Copeland, E. B. 1917. Keys to the ferns of Borneo. Sarawak Mus. J. 2: 301. Holttum, R. E. 1968. Revised Flora of Malaya. Vol. 2, Ferns. Ed. 2: 43–45.

18.1.1. Angiopteris angustifolia C. Presl, Suppl., 21 (1845). C. Chr. & Holttum, Gardens' Bull. 7: 209 (1934).

Terrestrial. Lower montane forest. Elevation: 1500 m.

Material examined: MESILAU RIVER: 1500 m, *RSNB 4128* (K); MINITINDUK: *Holttum SFN 25716* (SING).

18.1.2. Angiopteris brooksii Copel., Philipp. J. Sci. 10C: 145, t. 1 (1915). C. Chr. & Holttum, Gardens' Bull. 7: 209 (1934).

Terrestrial. Hill forest. Elevation: 900 m.

Material examined: DALLAS: 900 m, *Clemens 26825* (K), 900 m, *26826* (SING), 900 m, *27354* (K, SING).

18.1.3. Angiopteris evecta (G. Forster) Hoffm., Comm. Soc. Reg. Gott. 12: 29, t. 5 (1796).

Terrestrial. Lowland dipterocarp forest. Elevation: 500 m.

Material examined: NALUMAD: 500 m, *Shea & Aban SAN 77313* (K); PORING HOT SPRINGS: 500 m, *Parris & Croxall 8990* (K).

18.1.4. Angiopteris ferox Copel., Philipp. J. Sci. 6C: 134 (1911). C. Chr. & Holttum, Gardens' Bull. 7: 209 (1934).

Terrestrial. Lower montane forest. Elevation: 900–2900 m.

Material examined: KILEMBUN BASIN: 1100 m, *Clemens 34468* (K); KILEMBUN RIVER: 1500–2900 m, *Clemens 33865* (K); LIWAGU RIVER TRAIL: 1400 m, *Parris & Croxall 9034* (K); LIWAGU/MESILAU RIVERS: 1500 m, *RSNB 2812* (K, SING); TAHUBANG RIVER: 900 m, *Haviland 1476* (K); TENOMPOK: 1400 m, *Holttum SFN 25382* (K, SING), 1500 m, *Meijer SAN 20314* (K).

18.1.5. Angiopteris holttumii C. Chr. in C. Chr. & Holttum, Gardens' Bull. 7: 209 (1934). Type: Lubang: 1200 m, *Holttum SFN 25552* (holotype BM n.v.; isotypes K!, SING!).

Terrestrial. Hill forest and lower montane forest. Elevation: 1200–1800 m. Not endemic to Mount Kinabalu.

Additional material examined: KIAU: *Clemens 10231* (K); MAMUT ROAD: 1200–1300 m, *Tamura & Hotta 356* (K); PENIBUKAN: 1800 m, *Clemens 40983* (K).

18.2. CHRISTENSENIA

Holttum, R. E. 1968. Revised Flora of Malaya. Vol. 2, Ferns. Ed. 2: 45–46.

18.2.1. Christensenia aesculifolia (Blume) Maxon, Proc. Biol. Soc. Washington 18: 240 (1905).

Terrestrial. Lower montane forest? Elevation: 1400 m.

Material examined: KINATEKI RIVER: 1400 m, *Clemens 50410* (K).

18.3. MARATTIA

Copeland, E. B. 1958. Fern Flora of the Philippines. Vol. 1: 25–27.

18.3.1. Marattia sylvatica Blume, Enum. Pl. Javae, 256 (1828).

Marattia pellucida sensu C. Chr. & Holttum non C. Presl, Gardens' Bull. 7: 209 (1934).

Terrestrial. Hill forest and lower montane forest. Elevation: 1000–1500 m.

Material examined: EASTERN SHOULDER: 1200 m, *RSNB 1532* (K, SING); KIAU/TAHUBANG RIVER: 1000 m, *Parris 11533* (K); PENIBUKAN: 1200 m, *Clemens 32066* (K); TENOMPOK: 1500 m, *Clemens 27428* (K), 1500 m, *30459* (K), 1400 m, *Holttum SFN 25391* (K, SING).

19. MATONIACEAE

19.1. MATONIA

Holttum, R. E. 1968. Revised Flora of Malaya. Vol. 2, Ferns. Ed. 2: 58–60.

19.1.1. Matonia pectinata R. Br. in Wall., Pl. As. Rar. 1: t. 16 (1829).

Matonia foxworthyi Copel., Philipp. J. Sci. 3C: 343, pl. 2 (1908). C. Chr. & Holttum, Gardens' Bull. 7: 223 (1934).

Terrestrial. Lower montane forest in fairly open situations. Elevation: 1500–2100 m.

Material examined: KEMBURONGOH: 2100 m, *Clemens 28460* (K); MARAI PARAI: 1500 m, *Holttum SFN 25619* (K, SING).

20. OLEANDRACEAE

20.1. ARTHROPTERIS

Holttum, R. E. 1966. The genus *Arthropteris* J. Sm. in Malesia. Blumea 14: 225–229.

20.1.1. Arthropteris repens (Brack.) C. Chr., Bishop Mus. Bull. 177: 48 (1943).

Arthropteris obliterata sensu C. Chr. & Holttum non (R. Br.) J. Sm., Gardens' Bull. 7: 239 (1934).

Terrestrial climber. Hill forest to lower montane forest. Elevation: 1200–1500 m.

Material examined: DALLAS/TENOMPOK: 1200 m, *Clemens 27383* (SING); TENOMPOK: 1500 m, *Clemens 29568* (L, SING).

20.2. NEPHROLEPIS

Holttum, R. E. 1968. Revised Flora of Malaya. Vol. 2, Ferns. Ed. 2: 375–383.

20.2.1. Nephrolepis biserrata (Sw.) Schott agg., Gen. Fil. ad t. 3 (1834).

Terrestrial. Lowland forest. Elevation: 500 m.

Material examined: PORING HOT SPRINGS: 500 m, *Parris & Croxall 8996* (K); TAKUTAN: 500 m, *Shea & Aban SAN 77189* (K).

20.2.2. Nephrolepis cordifolia (L.) C. Presl, Tent., 79 (1836).

Terrestrial. Lower montane forest. Elevation: 1500 m.

Material examined: BAMBANGAN RIVER: 1500 m, *RSNB 4416* (K, SING), 1500 m, *RSNB 1296* (K, SING).

20.2.3. Nephrolepis davallioides (Sw.) Kunze, Bot. Zeit. 4: 460 (1846).

Nephrolepis acuminata (Houtt.) Kuhn, Ann. Lugd. Bat. 4: 286 (1869). C. Chr. & Holttum, Gardens' Bull. 7: 239 (1934).

Terrestrial. Hill forest to lower montane forest. Elevation: 800–1500 m.

Material examined: BUNDU TUHAN: 800 m, *Gibbs 3953* (K); DALLAS: 900 m, *Clemens 26925* (K), 900 m, *27353* (K); EASTERN SHOULDER: 1200 m, *RSNB 260* (K, SING); GURULAU SPUR: *Topping 1630* (SING, US); LIWAGU RIVER TRAIL: 1500 m, *Parris & Croxall 9030* (K), 1400 m, *9031* (K); MARAI PARAI: 1200 m, *Clemens 32389* (K); PENATARAN RIVER: 900 m, *Clemens 32569* (K); TENOMPOK: 1500 m, *Clemens 28065* (K, SING).

20.2.4. Nephrolepis falcata (Cav.) C. Chr., Dansk Bot. Arkiv 9: 15, t. 1 (5–9) (1937).

Nephrolepis barbata Copel., Perkins, Fragm., 178 (1905). C. Chr. & Holttum, Gardens' Bull. 7: 239 (1934).

Terrestrial and epiphytic. Hill forest, sometimes on ultramafics. Elevation: 700–1500 m.

Material examined: LOHAN RIVER: 700–900 m, *Beaman 9204* (MICH); MINITINDUK GORGE: 1500 m, *Clemens 29548* (SING).

20.2.5. Nephrolepis hirsutula (G. Forster) C. Presl, Tent., 79 (1836). C. Chr. & Holttum, Gardens' Bull. 7: 239 (1934).

Terrestrial. Lowland and lower montane forest. Elevation: 500–1500 m.

Material examined: KUNDASANG: 900 m, *Clemens 29241* (US); PARK HEADQUARTERS: 1500 m, *Parris & Croxall 9057* (K); PORING HOT SPRINGS: 500 m, *Parris & Croxall 8997* (K).

20.2.6. Nephrolepis radicans (N. L. Burm.) Kuhn, Ann. Lugd. Bat. 4: 285 (1869). C. Chr. & Holttum, Gardens' Bull. 7: 239 (1934).

Terrestrial. Hill forest. Elevation: 900 m.

Material examined: DALLAS: 900 m, *Clemens 27297* (K).

20.3. OLEANDRA

Copeland, E. B. 1917. Keys to the ferns of Borneo. Sarawak Mus. J. 2: 338–339. Holttum, R. E. 1968. Revised Flora of Malaya. Vol. 2, Ferns. Ed. 2: 383–387.

20.3.1. Oleandra oblanceolata Copel., Philipp. J. Sci. 7C: 64 (1912). C. Chr. & Holttum, Gardens' Bull. 7: 239 (1934) p.p.

Epiphytic, terrestrial, and scandent. Lower montane forest, hill forest. Elevation: 1200–2100 m.

Material examined: EASTERN SHOULDER: 1200 m, *RSNB 298* (K, SING); GOLF COURSE SITE: 1700–2000 m, *Beaman 9605* (MICH, MSC); KIAU VIEW TRAIL: 1600 m, *Parris 10800* (K), 1600 m, *Parris & Croxall 8551* (K); KUNDASANG: 1400 m, *RSNB 1461* (K, SING); LIWAGU RIVER TRAIL: 1500 m, *Parris 10824* (K); LUMU-LUMU: 2100 m, *Clemens 27210* (K); MAMUT COPPER MINE: 1600–1700 m, *Beaman 9921* (K, MICH, MSC), 1400–1500 m, *10332* (K, MICH, MSC); MAMUT RIVER: 1400 m, *RSNB 1273* (K, SING); MARAI PARAI SPUR: *Clemens 11053* (K); PENIBUKAN: 1500 m, *Clemens 30902* (K); TAHUBANG RIVER: 1200 m, *Clemens 30705* (K); TENOMPOK: 1500 m, *Clemens 28337* (K), 1500 m, *28763* (K).

20.3.2. Oleandra pistillaris (Sw.) C. Chr., Index Fil. Suppl. 3: 132 (1934).

Oleandra oblanceolata sensu C. Chr. & Holttum p.p. non Copel., Gardens' Bull. 7: 239 (1934).

Terrestrial. Lower montane forest. Elevation: 1500 m.

Material examined: LUMU-LUMU: 1500 m, *Holttum SFN 25723* (SING); MARAI PARAI SPUR: *Topping 1856* (SING).

20.3.3. Oleandra tricholepis Kunze, Bot. Zeit. 9: 349 (1851).

Terrestrial. Lower montane forest, apparently mostly on ultramafics. Elevation: 1800–2400 m.

Material examined: MARAI PARAI: 1800 m, *Clemens 32871* (K); PENATARAN RIVER: 2400 m, *Clemens 32552* (K); SUMMIT TRAIL: 1800 m, *Parris & Croxall 8873* (K).

21. OPHIOGLOSSACEAE

21.1. BOTRYCHIUM

Copeland, E. B. 1958. Fern Flora of the Philippines. Vol. 1: 19–20. Holttum, R. E. 1968. Revised Flora of Malaya. Vol. 2, Ferns. Ed. 2: 629.

21.1.1. Botrychium daucifolium Wall. ex Hook. & Grev., Ic. Fil., t. 161 (1829). C. Chr. & Holttum, Gardens' Bull. 7: 208 (1934).

Terrestrial. Lower montane forest. Elevation: 1500 m.

Material examined: MESILAU RIVER: 1500 m, *RSNB 4988* (K); TENOMPOK: 1500 m, *Clemens 27892* (K), 1500 m, *29210* (K).

21.2. HELMINTHOSTACHYS

21.2.1. Helminthostachys zeylanica (L.) Hook., Gen., t. 47 (1840). C. Chr. & Holttum, Gardens' Bull. 7: 209 (1934).

Terrestrial. Lowland and hill forest. Elevation: 600–900 m.

Material examined: DALLAS: 900 m, *Clemens 27666* (K); PORING HOT SPRINGS: 600 m, *Beaman 7561* (MICH, MSC).

21.3. OPHIOGLOSSUM

21.3.1. Ophioglossum intermedium Hook., Ic. Pl. t. 995 (1854). C. Chr. & Holttum, Gardens' Bull. 7: 208 (1934).

Terrestrial. Hill forest? Based on *Clemens 10243* (BM? n.v.).

21.3.2. Ophioglossum pendulum L., Sp. Pl. ed. 2, 2: 1518 (1763).

a. f. pendulum

Ophioglossum moultonii sensu C. Chr. & Holttum non Copel., Gardens' Bull. 7: 208 (1934).

Terrestrial. Lower montane forest. Elevation: 1500 m.

Material examined: TENOMPOK: 1500 m, *Clemens 28320* (K).

b. f. nutans Alderw., Handb. Malayan Ferns Suppl. 1: 454 (1917).

Terrestrial. Lower montane forest. Elevation: 1500–1600 m.

Material examined: MEMPENING TRAIL: 1600 m, *Parris & Croxall 9121* (K); MESILAU RIVER: 1500 m, *RSNB 4088* (K); PENATARAN RIVER: 1500 m, *Clemens 32527* (K); TENOMPOK: 1500 m, *Clemens 27468* (K).

21.3.3. Ophioglossum reticulatum L.

a. f. complicatum (Miq.) Wieffering, Blumea 12: 330 (1964).

Ophioglossum pedunculosum sensu C. Chr. & Holttum non Desv., Gardens' Bull. 7: 208 (1934).

Terrestrial. Hill forest to lower montane forest. Elevation: 900–2100 m.

Material examined: DALLAS: 900 m, *Clemens 26158* (K), 900 m, *Holttum SFN 25269* (K, SING); KIAU: *Clemens 10093* (K), *Topping 1636* (SING); MAMUT RIVER: 1200 m, *RSNB 1231* (K), 1200 m, *1245* (K), 1200 m, *1257* (K); MARAI PARAI: *Topping 1889* (SING); MESILAU BASIN: 2100 m, *Clemens 29280* (K); MESILAU RIVER: 2000 m, *Collenette 21628* (K); PARK HEADQUARTERS: 1500 m, *Parris 11646* (K).

22. PLAGIOGYRIACEAE

22.1. PLAGIOGYRIA

Copeland, E. B. 1929. The fern genus *Plagiogyria*. Philipp. J. Sci. 38: 377–417.

22.1.1. Plagiogyria adnata (Blume) Bedd., Ferns Br. India, t. 51 (1865). C. Chr. & Holttum, Gardens' Bull. 7: 225 (1934).

Terrestrial. Hill forest and lower montane forest. Elevation: 900–2100 m.

Material examined: DALLAS: 900 m, *Clemens 27960* (K); KEMBURONGOH: 2100 m, *Sinclair et al. 9048* (K, SING); KIAU VIEW TRAIL: 1600 m, *Parris & Croxall 8544* (K); TENOMPOK/LUMU-LUMU: 1500 m, *Holttum SFN 25442* (K, SING).

22.1.2. Plagiogyria clemensiae Copel., Philipp. J. Sci. 38: 395, pl. 4 (1929). C. Chr. & Holttum, Gardens' Bull. 7: 225 (1934). Type: Paka-paka Cave: *Clemens 10589* (holotype PNH?†; isotype BM n.v.).

Terrestrial. Lower montane and upper montane forest, sometimes on ultramafics. Elevation: 1700–3800 m. Not endemic to Mount Kinabalu.

Material examined: DACHANG: 3000 m, *Clemens 29059* (K, SING); EASTERN SHOULDER: 3000 m, *RSNB 729* (K, SING); KEMBURONGOH: 2400 m, *Clemens 28962* (K), 2400 m, *28992* (SING); KINATEKI RIVER HEAD: 2700 m, *Clemens 31822* (US), 2100–2700 m, *31989* (K); LAYANG-LAYANG/PAKA-PAKA CAVE: 3000 m, *Edwards 2189* (K); LIWAGU RIVER TRAIL: 1800 m, *Parris 11466* (K); MT. NUNGKEK: 1700 m, *Clemens 32636* (K); PAKA-PAKA CAVE: 3000 m, *Holttum SFN 25730* (SING), 3000 m, *Sinclair et al. 9121* (K, SING); SHEILA'S PLATEAU/SHANGRI LA VALLEY: 3400 m, *Collenette 21517* (K); SUMMIT AREA: 3700 m, *Clemens 28961* (K); SUMMIT TRAIL: 2400 m, *Parris 11541* (K), 2700 m, *11565* (K), 3200 m, *Parris & Croxall 8746* (K); UPPER KINABALU: 3000–3800 m, *Clemens 51112* (K).

22.1.3. Plagiogyria glauca (Blume) Mett., Plagiog., 9, no. 3 (1858). C. Chr. & Holttum, Gardens' Bull. 7: 225 (1934).

Terrestrial. Upper montane forest. Elevation: 2100–3700 m.

Material examined: DACHANG: 3000 m, *Clemens 29060* (K); GOKING'S VALLEY: 2800 m, *Fuchs 21488* (K); KEMBURONGOH: 2100 m, *Holttum s.n.* (SING); PAKA-PAKA CAVE: 3500–3700 m, *Molesworth-Allen 3270* (US); SUMMIT AREA: 3700 m, *Clemens 28385* (K, SING); SUMMIT TRAIL: 3200 m, *Parris & Croxall 8745* (K).

22.1.4. Plagiogyria pycnophylla (Kunze) Mett., Plagiog., 8, no. 2 (1858). C. Chr. & Holttum, Gardens' Bull. 7: 224 (1934).

Terrestrial. Upper montane forest and summit area. Elevation: 3300–4100 m.

Material examined: LOW'S PEAK: 4100 m, *Clemens 27043* (K); MOUNT KINABALU: 4000 m, *Holttum SFN 25491* (K, SING); PAKA-PAKA CAVE: *Clemens 10590* (K); SUMMIT AREA: *Clemens 28960* (K); SUMMIT TRAIL: 3300 m, *Parris & Croxall 8747* (K); VICTORIA PEAK: 3700 m, *Clemens s.n.* (SING).

23. POLYPODIACEAE

23.1. AGLAOMORPHA

Roos, M. C. 1985. Phylogenetic systematics of the Drynarioideae. Thesis, Leiden, University Printer.

23.1.1. Aglaomorpha brooksii Copel., Philipp. J. Sci. 6C: 141, t. 25 (1911).

Aglaomorpha splendens sensu C. Chr. & Holttum non (J. Sm.) Copel., Gardens' Bull. 7: 314 (1934).

Epiphyte. Hill dipterocarp and lower montane forest. Elevation: 1000–1500 m.

Material examined: KIAU/LUBANG: *Topping 1805* (SING); PENIBUKAN: 1200 m, *Clemens 32122* (K); PINOSUK PLATEAU: 1500 m, *Parris & Croxall 9155* (K); PORING HOT SPRINGS/LANGANAN WATER FALLS: 1000 m, *Parris & Croxall 9178* (K); TENOMPOK: 1500 m, *Clemens 29038* (K, SING).

23.2. BELVISIA

Christensen, C. 1929. Taxonomic fern-studies I. Revision of the polypodioid genera with longitudinal coenosori (Cochlidiinae and "Drymoglossinae"); with a discussion of their phylogeny. Dansk Bot. Arkiv 6: 54–70 [as *Hymenolepis*].

23.2.1. Belvisia mucronata (Fée) Copel., Gen. Fil, 192 (1947).

Hymenolepis revoluta Blume var. *planiuscula* sensu C. Chr. & Holttum non (Mett.) Hieron. ex C. Chr., Gardens' Bull. 7: 313 (1934) p.p.

Epiphyte. Lowland dipterocarp forest to lower montane forest. Elevation: 500–1500 m.

Material examined: DALLAS/TENOMPOK: 1200 m, *Holttum SFN 25292* (K, SING); KUNDASANG: *Clemens 29106* (K); PORING HOT SPRINGS: 500 m, *Parris & Croxall 8567* (K); TAHUBANG RIVER: 900 m, *Clemens 33230* (K); TENOMPOK: 1500 m, *Clemens 26975* (K), 1500 m, *28191* (K).

23.2.2. Belvisia spicata (L. f.) Mirb. ex Copel., Gen. Fil., 192 (1947).

Hymenolepis revoluta Blume, Enum. Pl. Javae, 201 (1828). C. Chr. & Holttum, Gardens' Bull. 7: 313 (1934). *Hymenolepis revoluta* Blume var. *planiuscula* (Mett.) Hieron. ex C. Chr., Dansk Bot. Arkiv 6: 58 (1929). C. Chr. & Holttum, Gardens' Bull. 7: 313 (1934) p.p.

Epiphyte. Hill forest and lower montane forest. Elevation: 500–2200 m.

Material examined: BUNDU TUHAN: 1200 m, *Beaman 10527* (K, MICH, MSC); DALLAS: 900 m, *Clemens 27503* (US); EAST MESILAU/MENTAKI RIVERS: 1700 m, *Beaman 9373* (K, MICH, MSC); GOLF COURSE SITE: 1700 m, *Beaman 8498* (MICH), 1700 m, *8555* (MICH); KUNDASANG: 900 m, *Clemens 29106* (US); MESILAU CAVE: 1900–2200 m, *Beaman 9564* (K, MICH, MSC); PARK HEADQUARTERS: 1500 m, *Parris & Croxall 8563* (K); PENATARAN RIVER: 500 m, *Beaman 9321* (MICH); PINOSUK PLATEAU: 1400 m, *Beaman 10736* (MICH); TENOMPOK: 1500 m, *Clemens 26975* (US); WEST MESILAU RIVER: 1600–1700 m, *Beaman 8651* (MICH, MSC).

23.2.3. Belvisia squamata (Hieron. ex C. Chr.) Copel. Gen. Fil., 192 (1947).

a. var. borneensis (C. Chr.) Parris, comb. nov.

Hymenolepis squamata Hieron. ex C. Chr. var. *borneensis* C. Chr., Dansk Bot. Arkiv 6: 60 (1929). C. Chr. & Holttum, Gardens' Bull. 7: 313 (1934). Type: Summit Trail: *Clemens 10664* (holotype PNH?†).

Lithophyte and epiphyte. Tall and short upper montane forest. Elevation: 1800–3800 m. Endemic to Mount Kinabalu.

Material examined: EASTERN SHOULDER: 2900 m, *RSNB 804* (K, SING); KILEMBUN RIVER: 2700 m, *Clemens 33737* (K); LUMU-LUMU: 2100 m, *Clemens 27955* (K); MESILAU RIVER: 1800 m, *Clemens 51364* (K); MOUNT KINABALU: 3500 m, *Holttum SFN 25487* (K, SING); PAKA-PAKA CAVE: 3000 m, *Molesworth-Allen 3250* (K), 3100 m, *Parris 11575* (K), 3000 m, *Parris & Croxall 8714* (K), *Topping 1675* (US); PINOSUK PLATEAU: 2900 m, *RSNB 794b* (K, SING); SUMMIT AREA: 3700 m, *Clemens 27048* (K), 3300 m, *Sinclair et al. 9154* (K, SING); SUMMIT TRAIL: 3300 m, *Parris & Croxall 8767* (K), *Topping 1703* (US); VICTORIA PEAK: 3800 m, *Clemens 51396* (K).

23.3. COLYSIS

Holttum, R. E. 1968. Revised Flora of Malaya. Vol. 2, Ferns. Ed. 2: 159–163.

23.3.1. Colysis acuminata Holttum, Rev. Fl. Malaya 2: 162, f. 73 (1955).

Gymnogramme acuminata Baker non Kaulf. J. Bot. 26: 326 (1888). *Polypodium interruptum* C. Chr., Index Fil., 333 (1905). C. Chr. & Holttum, Gardens' Bull. 7: 308 (1934).

Lithophyte. Lower montane forest. Elevation: 900–2100 m.

Material examined: LUBANG: *Topping 1774* (SING, US); LUMU-LUMU: 2100 m, *Clemens 29968* (US); MINITINDUK GORGE: 900–1200 m, *Clemens 29624* (K, SING); MOUNT KINABALU: 1500 m, *Holttum 39* (SING); PINOSUK PLATEAU: 1500 m, *Holttum 39* (K); SUMMIT TRAIL: 1800 m, *Parris & Croxall 8884* (K); TENOMPOK: 1600 m, *Clemens 29312* (K), 1200 m, *Holttum SFN 25300* (K, SING); TIBABAR FALLS: 1800 m, *Parris 11463* (K).

23.3.2. Colysis loxogrammoides (Copel.) M. G. Price, Contr. Univ. Mich. Herb. 16: 193 (1987).

Polypodium polysorum Brause, Bot. Jahrb. 56: 203 (1920). C. Chr. & Holttum, Gardens' Bull. 7: 308 (1934).

Terrestrial. Lower montane forest. Elevation: 1200–1500 m.

Material examined: KUNDASANG: 1200 m, *RSNB 1432* (K); TENOMPOK: 1500 m, *Clemens 26184* (K, SING), 1500 m, *28642* (SING).

23.3.3. Colysis macrophylla (Blume) C. Presl, Epim., 147 (1849).

Lithophyte and epiphyte low on trees. Lowland dipterocarp forest and lower montane forest. Elevation: 500–1400 m.

Material examined: KIAU: *Topping 1572* (US); KIPUNGIT FALLS: 500 m, *Parris & Croxall 8900* (K); KUNDASANG: 1200 m, *RSNB 1420* (SING), 1200 m, *1420a* (K); LIWAGU RIVER TRAIL: 1400 m, *Parris & Croxall 8605* (K).

23.3.4. Colysis pedunculata (Hook. & Grev.) Ching, Bull. Fan Mem. Inst. Biol. Bot. 4: 321 (1933).

Lithophyte. Lowland dipterocarp forest. Elevation: 600 m.

Material examined: PORING HOT SPRINGS/LANGANAN WATER FALLS: 600 m, *Parris & Croxall 8966* (K).

23.4. CRYPSINUS

Parris, B. S., Jermy, A. C., Camus, J. M. & Paul, A. M. 1984. The Pteridophyta of Gunung Mulu National Park. *In* Studies on the Flora of Gunung Mulu National Park, Sarawak. Ed. A. C. Jermy: 195. Forest Dept., Kuching, Sarawak.

23.4.1. Crypsinus albidopaleatus (Copel.) Copel., Gen. Fil., 207 (1947).

Polypodium albidopaleatum Copel., Philipp. J. Sci. 12C: 63 (1917). C. Chr. & Holttum, Gardens' Bull. 7: 307 (1934). Type: Lubang/Paka-paka Cave: *Topping 1749* (holotype PNH?†; isotypes (MICH!, SING!, US!)

Epiphyte. Lower montane forest. Elevation: 1200–2900 m. Not endemic to Mount Kinabalu.

Additional material examined: EASTERN SHOULDER: 2400 m, *RSNB 183* (K, SING); GURULAU SPUR: 2100–2400 m, *Clemens 51040* (K); KEMBURONGOH: 2100 m, *Clemens 28387* (K), 2100 m, *Sinclair et al. 9055A* (SING), 2100 m, *9055a* (K); KEMBURONGOH/LUMU-LUMU: 1800 m, *Holttum SFN 25452* (K, SING); KILEMBUN BASIN: 2900 m, *Clemens 33681* (K); MAMUT RIVER: 1200 m, *RSNB 1240* (SING); SUMMIT TRAIL: 2600 m, *Parris 11550* (K), 2100 m, *Parris & Croxall 8690* (K).

23.4.2. Crypsinus enervis (Cav.) Copel., Gen. Fil., 207 (1947).

Polypodium rupestre Blume, Enum. Pl. Javae, 124 (1828). C. Chr. & Holttum, Gardens' Bull. 7: 307 (1934).

Epiphyte. Lower montane forest. Elevation: 1300–2900 m.

Material examined: EASTERN SHOULDER, CAMP 4: 2900 m, *RSNB 1159* (K, SING); LIWAGU RIVER TRAIL: 1700 m, *Parris 11659* (K); MESILAU BASIN: 2100 m, *Clemens 29030* (K, SING), 2100–2400 m, *29705* (K, SING); MOUNT KINABALU: 1300–1400 m, *Kodama 12691* (KYO); PARK HEADQUARTERS: 1500 m, *Parris & Croxall 8564* (K), 1500 m, *8572* (K); PINOSUK PLATEAU: 1400 m, *Beaman 10733* (MICH), 1400 m, *10735* (MICH), 1400 m, *10751* (MICH); SUMMIT TRAIL: 2100 m, *Parris & Croxall 8668* (K); TENOMPOK: 1500 m, *Clemens 27423* (K); TENOMPOK/LUMU-LUMU: 1500 m, *Holttum SFN 25441* (SING).

23.4.3. Crypsinus platyphyllus (Sw.) Copel., Gen. Fil., 207 (1947).

Polypodium platyphyllum Sw., Syn., 27 (1806). C. Chr. & Holttum, Gardens' Bull. 7: 308 (1934).

Epiphyte. Lower montane forest. Elevation: 1500 m.

Material examined: TENOMPOK: 1500 m, *Clemens 29545* (K, SING).

23.4.4. Crypsinus stenophyllus (Blume) Holttum, Rev. Fl. Malaya 2: 199, f. 101 (1955).

Polypodium stenophyllum Blume, Enum. Pl. Javae, 124 (1828). C. Chr. & Holttum, Gardens' Bull. 7: 307 (1934).

Epiphyte. Upper montane forest. Elevation: 2900–3200 m.

Material examined: EASTERN SHOULDER: 3200 m, *RSNB 854* (K, SING); EASTERN SHOULDER, CAMP 4: 2900 m, *RSNB 1157* (K, SING); MOUNT KINABALU: 3200 m, *Haviland 1489* (K).

23.4.5. Crypsinus stenopteris (Baker) Parris, Fern Gaz. 12: 118 (1980).

Polypodium stenopteris Baker, J. Bot. 17: 43 (1879). C. Chr. & Holttum, Gardens' Bull. 7: 307 (1934).

Epiphyte and lithophyte. Lower and upper montane forest, sometimes on ultramafics. Elevation: 1500–3000 m.

Material examined: DACHANG: 2100–3000 m, *Clemens 29277* (K); EASTERN SHOULDER: 2400 m, *RSNB 181* (K, SING); EASTERN SHOULDER, CAMP 4: 2700 m, *RSNB 1140* (K, SING); GURULAU SPUR: 2400 m, *Clemens 50919* (K); JANET'S HALT/SHEILA'S PLATEAU: 2900 m, *Collenette 21528* (K); KEMBURONGOH: 2400–2700 m, *Clemens 27161* (K), 2400 m, *29277b* (K), 2300 m, *Sinclair et al. 9078* (K); KEMBURONGOH/LAYANG-LAYANG: 2400 m, *Edwards 2203* (K); KEMBURONGOH/LUMU-LUMU: 1800 m, *Holttum SFN 25453* (K, SING); KEMBURONGOH/PAKA-PAKA CAVE:

2300 m, *Sinclair et al. 9078* (SING), *Topping 1656* (SING, US); KILEMBUN RIVER HEAD: 2400–2700 m, *Clemens 31740* (K); LUBANG/PAKA-PAKA CAVE: *Topping 1734* (US); LUMU-LUMU: 1800 m, *Clemens 27063* (K); MAMUT COPPER MINE: 1600–1700 m, *Beaman 9927* (K, MICH, MSC); MAMUT RIVER: 1500 m, *RSNB 1270* (K, SING); MARAI PARAI: 2400–2700 m, *Clemens s.n.* (US); MOUNT KINABALU: *Haslam s.n.* (US); PIG HILL: 2000–2300 m, *Beaman 9867* (K, MICH, MSC); SUMMIT TRAIL: 2600 m, *Parris 11554* (K), 2600 m, *Parris & Croxall 8808* (K).

23.4.6. Crypsinus taeniophyllus (Copel.) Copel., Gen. Fil., 207 (1947).

Polypodium taeniophyllum Copel., Philipp. J. Sci. 7C: 65: (1912). C. Chr. & Holttum, Gardens' Bull. 7: 307 (1934).

Material examined: DALLAS: *Clemens s.n.* (BM).

23.5. DRYMOGLOSSUM

Holttum, R. E. 1968. Revised Flora of Malaya. Vol. 2, Ferns. Ed. 2: 149–150.

23.5.1. Drymoglossum piloselloides (L.) C. Presl, Tent., 227, t. 10, f. 5, 6 (1836). C. Chr. & Holttum, Gardens' Bull. 7: 314 (1934).

Epiphyte. Lowland forest. Elevation: 500–900 m.

Material examined: DALLAS: 900 m, *Clemens 27296* (K); TAKUTAN: 500 m, *Shea & Aban SAN 77133* (K).

23.6. DRYNARIA

Roos, M. C. 1985. Phylogenetic systematics of the Drynarioideae (Polypodiaceae). Thesis, Leiden, University Printer.

23.6.1. Drynaria quercifolia (L.) J. Sm., J. Bot. 3: 398 (1841).

Epiphyte. Lowland dipterocarp and hill forest. Elevation: 500–1400 m.

Material examined: KIPUNGIT FALLS: 500 m, *Parris & Croxall 8908* (K); LOHAN RIVER: 700–900 m, *Beaman 9237* (MICH); LOHAN/MAMUT COPPER MINE: 900 m, *Beaman 10603* (K, MICH, MSC); PINAWANTAI: 1400 m, *Shea & Aban SAN 76864* (K).

23.6.2. Drynaria sparsisora (Desv.) T. Moore, Index, 348 (1862). C. Chr. & Holttum, Gardens' Bull. 7: 315 (1934).

Epiphyte. Lower montane forest. Elevation: 1500 m.

Material examined: TENOMPOK: 1500 m, *Clemens 28684* (K, SING).

23.7. DRYNARIOPSIS

Roos, M. C. 1985. Phylogenetic systematics of the Drynarioideae (Polypodiaceae). Thesis, Leiden, University Printer; under *Aglaomorpha*.

23.7.1. Drynariopsis heraclea (Kunze) Ching, Sunyatsenia 5: 262 (1940).

Polypodium heracleum Kunze, Bot. Zeit. 6: 117 (1848). C. Chr. & Holttum, Gardens' Bull. 7: 311 (1934).

Epiphyte. Lowland dipterocarp forest to lower montane forest. Elevation: 500–1500 m.

Material examined: MINITINDUK: 900 m, *Holttum SFN 25593* (K, SING); PORING HOT SPRINGS: 500 m, *Parris & Croxall 9005* (K); TENOMPOK: 1500 m, *Clemens 29668* (K).

23.8. GONIOPHLEBIUM

Holttum, R. E. 1968. Revised Flora of Malaya. Vol. 2, Ferns. Ed. 2: 202–207 [as *Polypodium*]. Parris, B. S., Jermy, A. C., Camus, J. M., & Paul, A. M. 1984. The Pteridophyta of Gunung Mulu National Park. *In* Studies on the Flora of Gunung Mulu National Park, Sarawak. Ed. A. C. Jermy: 197–198. Forest Dept., Kuching, Sarawak. Rödl-Linder, G. 1990. A monograph of the fern genus *Goniophlebium* (Polypodiaceae). Blumea 34: 277–423.

23.8.1. Goniophlebium percussum (Cav.) W. H. Wagner & Grether, Occ. Papers Bishop Mus. 19: 88 (1948).

Polypodium verrucosum Wall. ex Hook., Garden Ferns, t. 41 (1862). C. Chr. & Holttum, Gardens' Bull. 7: 306 (1934).

Epiphyte. Hill forest? Elevation: 1200 m.

Material examined: DALLAS: 1200 m, *Clemens 27350* (K).

23.8.2. Goniophlebium persicifolium (Desv.) Bedd., Ferns Br. India correct. (1870).

Polypodium integrius Copel. ('*integriore*'), Philipp. J. Sci. 2C: 139 (1907). C. Chr. & Holttum, Gardens' Bull. 7: 305 (1934).

Epiphyte. Lower montane forest. Elevation: 1200–1500 m.

Material examined: PENIBUKAN: 1200–1500 m, *Clemens 51701* (K); TAHUBANG RIVER: 1200 m, *Clemens 31039* (K); TENOMPOK: *Clemens s.n.* (BM).

23.8.3. Goniophlebium rajaense (C. Chr.) Parris, Fern Gaz. 12: 118 (1980).

Polypodium rajaense C. Chr. in Irmsch., Mitt. Inst. Allg. Bot. Hamburg 7: 159 (1928). *Polypodium integriore* Copel. var. *rajaense* (C. Chr.) C. Chr. in C. Chr. & Holttum, Gardens' Bull. 7: 305 (1934).

Epiphyte. Lower montane forest. Elevation: 1400–1700 m.

Material examined: KILEMBUN RIVER HEAD: 1700 m, *Clemens 32447* (US); LIWAGU RIVER TRAIL: 1400 m, *Parris & Croxall 8602* (K); MOUNT KINABALU: 1500 m, *Shim SAN 75401* (K); TAHUBANG RIVER: *Clemens 30704* (K); TENOMPOK: 1500 m, *Clemens 26855* (US), 1500 m, *29484* (K), 1400 m, *Holttum SFN 25393* (SING, US).

23.8.4. Goniophlebium subauriculatum (Blume) C. Presl, Tent., 186 (1836).

Polypodium subauriculatum Blume, Enum. Pl. Javae, 133 (1828). C. Chr. & Holttum, Gardens' Bull. 7: 305 (1934). *Polypodium pallens* Blume, Fl. Javae 2: 178, t. 84, f. 1 (1829). C. Chr. & Holttum, Gardens' Bull. 7: 305 (1934).

Epiphyte and lithophyte. Lowland dipterocarp forest and lower montane forest. Elevation: 500–2400 m.

Material examined: DALLAS: 1200 m, *Clemens 27431* (K); KADAMAIAN FALLS TRAIL: 1700 m, *Parris & Croxall 9082* (K); KAUNG: *Topping 1895* (SING, US); KIAU: 900 m, *Holttum SFN 25594* (SING), *Topping 1511* (US); KILEMBUN RIVER: 2400 m, *Clemens 33746* (US); LUBANG: 1200 m, *Holttum SFN 25547* (K, SING); MINITINDUK: 900–1200 m, *Clemens 29627* (US); PORING HOT SPRINGS: 500 m, *Parris & Croxall 8989* (K); TENOMPOK: 1500 m, *Clemens 29625* (K).

23.9. LECANOPTERIS

Holttum, R. E. 1968. Revised Flora of Malaya. Vol. 2, Ferns. Ed. 2: 190, 208–210.

23.9.1. Lecanopteris pumila Blume ex Copel., Polyp. Philipp., 134 (1905).

Lecanopteris carnosa sensu C. Chr. & Holttum non (Reinw.) Blume, Gardens' Bull. 7: 314 (1934).

Epiphyte. Lower montane forest. Elevation: 1400–1500 m.

Material examined: GURULAU SPUR: 1500 m, *Clemens 50453* (K); PARK HEADQUARTERS: 1400 m, *Abbe et al. 9975* (K).

23.9.2. Lecanopteris sinuosa (Wall. ex Hook.) Copel., Gen. Fil., 205 (1947).

Polypodium sinuosum Wall. ex Hook., Sp. Fil. 5: 61, t. 284 (1863). C. Chr. & Holttum, Gardens' Bull. 7: 311 (1934).

Epiphyte. Lowland forest.

Material examined: NALUMAD: *Shea & Aban SAN 77265* (K).

23.10. LEMMAPHYLLUM

Holttum, R. E. 1968. Revised Flora of Malaya. Vol. 2, Ferns. Ed. 2: 152–153.

23.10.1. Lemmaphyllum accedens (Blume) Donk, Reinwardtia 2: 409 (1954).

Polypodium accedens Blume, Enum. Pl. Javae, 121 (1828). C. Chr. & Holttum, Gardens' Bull. 7: 307 (1934).

Epiphyte. Lowland forest and lower montane valley forest. Elevation: 200–1700 m.

Material examined: BAMBANGAN RIVER: 1500 m, *RSNB 12* (L), 1500 m, *1294* (K, SING); DALLAS: 900 m, *Clemens 28199* (K); KEBAYAU: 200 m, *Holttum s.n.* (SING); LIWAGU RIVER TRAIL: 1600 m, *Parris 11478* (K), 1400 m, *Parris & Croxall 8611* (K); PENATARAN BASIN: 1400 m, *Clemens 40176* (K); PINOSUK PLATEAU: 1600 m, *Beaman 10799* (K, MICH, MSC); TENOMPOK: 1500 m, *Clemens 26704* (K, L); WEST MESILAU RIVER: 1600–1700 m, *Beaman 8663* (K, MICH, MSC).

23.11. LOXOGRAMME

Copeland, E. B. 1960. Fern Flora of the Philippines. Vol. 3: 540–542.

23.11.1. Loxogramme carinata M. G. Price, Amer. Fern J. 80: 4, f. 1 (1990). Type: Mamut River: 1200 m, *RSNB 1228* (holotype L n.v.; isotype K!).

Loxogramme involuta C. Presl var. *gigas* sensu C. Chr. & Holttum non Copel., Gardens' Bull. 7: 312 (1934).

Epiphyte. Lower montane forest. Elevation: 800–1400 m. Not endemic to Mount Kinabalu.

Additional material examined: DALLAS: 800 m, *Clemens 26489* (BM), 900 m, *27300* (K); DALLAS/TENOMPOK: 900–1200 m, *Clemens 27037* (BM); PORING HOT SPRINGS: *Kokawa & Hotta 4869* (KYO); TENOMPOK: 1400 m, *Holttum SFN 25356b* (K).

23.11.2. Loxogramme ensifrons Alderw., Bull. Jard. Bot. Buit. 2, 11: 16, pl. 4 (1913). C. Chr. & Holttum, Gardens' Bull. 7: 311 (1934).

Epiphyte. Lowland dipterocarp and hill forest. Elevation: 500–1500 m.

Material examined: DALLAS: 900 m, *Clemens 27307* (K); DALLAS/TENOMPOK: 1200 m, *Holttum SFN 25291* (K); KIPUNGIT FALLS: 500 m, *Parris & Croxall 8520* (K); PORING HOT SPRINGS: 500 m, *Parris & Croxall 9016* (K); TENOMPOK: 1500 m, *Clemens 29229* (K).

23.11.3. Loxogramme nidiformis C. Chr. in C. Chr. & Holttum, Gardens' Bull. 7: 312 (1934). Type: Tenompok: *Holttum SFN 25356* (holotype BM n.v.; isotype K!).

Epiphyte. Lower montane valley forest. Elevation: 1400–1600 m. Not endemic to Mount Kinabalu.

Additional material examined: LIWAGU RIVER: 1400 m, *Parris & Croxall 8527* (K); LIWAGU RIVER TRAIL: 1600 m, *Parris 11480* (K), 1400 m, *Parris & Croxall 8527* (K); MESILAU RIVER: 1500 m, *RSNB 1377* (K); TENOMPOK: 1500 m, *Clemens 26949* (K).

23.11.4. Loxogramme parallela Copel., Philipp. J. Sci. 56: 106, t. 12 (1935). C. Chr. & Holttum, Gardens' Bull. 7: 311 (1934).

Loxogramme major Copel., Philipp. J. Sci. 56: 106, t. 12 (1935).

Epiphyte and lithophyte. Upper montane forest. Elevation: 2900–3100 m.

Material examined: EASTERN SHOULDER: 2900 m, *RSNB 801* (K); PAKA-PAKA CAVE: 3000 m, *Holttum SFN 25504* (K), 3100 m, *Parris 11576* (K); SUMMIT TRAIL: 3000 m, *Parris & Croxall 8508* (K).

23.11.5. Loxogramme wallichiana (Hook.) M. G. Price, Amer. Fern J. 74: 61 (1984).

Epiphyte and terrestrial. Lowland dipterocarp forest and lower montane valley forest. Elevation: 500–1700 m.

Material examined: PORING HOT SPRINGS: 500 m, *Parris & Croxall 9015* (K); WEST MESILAU RIVER: 1600–1700 m, *Beaman 8626* (K, MICH, MSC), 1600 m, *9358* (K, MICH, MSC).

23.12. MERINTHOSORUS

Roos, M. C. 1985. Phylogenetic systematics of the Drynarioideae (Polypodiaceae). Thesis, Leiden, University Printer; as *Aglaomorpha*.

23.12.1. Merinthosorus drynarioides (Hook.) Copel., Philipp. J. Sci. 6C: 92 (1911). C. Chr. & Holttum, Gardens' Bull. 7: 314 (1934).

Epiphyte. Lowland dipterocarp forest and lower montane forest. Elevation: 500–1500 m.

Material examined: KIAU: 900 m, *Holttum SFN 25624* (SING); PENATARAN RIVER: 500 m, *Beaman 9304* (MICH); PORING HOT SPRINGS: 500 m, *Parris & Croxall 9007* (K); TENOMPOK: 1500 m, *Clemens 29011* (K).

23.13. MICROSORUM

Copeland, E. B. 1960. Fern Flora of the Philippines. Vol. 3: 476–487. Holttum, R. E. 1968. Revised Flora of Malaya. Vol. 2, Ferns. Ed. 2: 170–180 [as *Microsorum*], 188-193 [as *Phymatodes*].

23.13.1. Microsorum commutatum (Blume) Copel., Gen. Fil., 196 (1947).

Polypodium commutatum Blume, Enum. Pl. Javae addend. (1828). C. Chr. & Holttum, Gardens' Bull. 7: 309 (1934).

Epiphyte and lithophyte. Lowland dipterocarp forest (apparently rarely in lower montane forest). Elevation: 500–1500 m.

Material examined: KADAMAIAN RIVER NEAR MINITINDUK: 900 m, *Holttum SFN 25572* (K); KIAU: 900 m, *Clemens 30480* (K), *Topping 1556* (SING, US); KIAU/LUBANG: *Topping 1802* (SING, US); MINITINDUK: 900 m, *Holttum SFN 25572* (SING); PORING HOT SPRINGS: 500 m, *Parris & Croxall 9009* (K); TENOMPOK: 1500 m, *Clemens 29582* (K).

23.13.2. Microsorum congregatifolium (Alderw.) Holttum, Rev. Fl. Malaya 2: 178 (1955).

Epiphyte. Lower montane forest? Elevation: 1200–1300 m.

Material examined: MAMUT ROAD: 1200–1300 m, *Tamura & Hotta 507* (K); PENIBUKAN: 1200 m, *Clemens 32033* (K).

23.13.3. Microsorum heterocarpum (Blume) Holttum, Rev. Fl. Malaya 2: 178, f. 87 (1955).

Polypodium heterocarpum (Blume) Mett., Fil. Lips., 37, t. 25, f. 24, 25 (1856). C. Chr. & Holttum, Gardens' Bull. 7: 308 (1934).

Lithophyte and epiphyte. Hill forest and lower montane forest. Elevation: 300–1700 m. The combination has been erroneously attributed to Ching.

Material examined: DALLAS: 900 m, *Clemens 26800* (K), 900 m, *27305* (K); LINISIHANG RIVER: 300 m, *Holttum SFN 25628* (K); LIWAGU RIVER TRAIL: 1700 m, *Parris & Croxall 9091* (K); TENOMPOK: 1400 m, *Holttum SFN 25720* (K, SING).

23.13.4. Microsorum insigne (Blume) Copel., Univ. Calif. Publ. Bot. 16: 112 (1929).

Polypodium insigne Blume, Enum. Pl. Javae, 127 (1828). C. Chr. & Holttum, Gardens' Bull. 7: 309 (1934).

Epiphyte and lithophyte. Lowland dipterocarp forest. Elevation: 500–900 m.

Material examined: DALLAS: 800 m, *Holttum SFN 25274* (K, SING); EASTERN SHOULDER: 800 m, *RSNB 577* (K, SING); EASTERN SHOULDER, CAMP 1: *RSNB 1183* (K, SING); KIPUNGIT FALLS: 500 m, *Parris & Croxall 8901* (K); LOHAN/MAMUT COPPER MINE: 900 m, *Beaman 10602* (K, MICH, MSC).

23.13.5. Microsorum ithycarpum (Copel.) Parris, comb. nov.

Polypodium ithycarpum Copel., Philipp. J. Sci. 12C: 64 (1917). C. Chr. & Holttum, Gardens' Bull. 7: 309 (1934). Type: Kiau: *Topping 1578* (holotype PNH?†).

Epiphyte. Hill forest and lower montane forest. Elevation: 900–1700 m. Endemic to Mount Kinabalu.

Material examined: DALLAS: 900 m, *Clemens 27025* (K), 900 m, *27294* (K), 900 m, *28387* (K); EAST MESILAU/MENTAKI RIVERS: 1700 m, *Beaman 9372* (K, MICH, MSC); LUBANG: *Topping 1778* (US); MINITINDUK: 900 m, *Holttum SFN 25568* (K, SING); PINOSUK PLATEAU: 1500 m, *Parris & Croxall 9142* (K).

23.13.6 Microsorum nigrescens (Blume) Copel., Occ. Papers Bishop Mus. 14: 74 (1938).

Ferns of Kinabalu

Epiphyte and lithophyte. Lowland and lower montane forest. Elevation: 500–1400 m.

Material examined: PENATARAN RIVER: 500 m, *Beaman 8872* (MICH, MSC); PINAWANTAI: 500 m, *Shea & Aban SAN 76905* (K); PINOSUK PLATEAU: 1400 m, *Beaman 10731* (MICH, MSC).

23.13.7. Microsorum punctatum (L.) Copel., Univ. Calif. Publ. Bot. 16: 111 (1929).

Epiphyte. Lowland dipterocarp forest. Elevation: 500 m.

Material examined: PORING HOT SPRINGS: 500 m, *Parris & Croxall 9013* (K).

23.13.8. Microsorum rubidum (Kunze) Copel., Gen. Fil., 197 (1947).

Epiphyte. Lower montane forest. Elevation: 1200 m.

Material examined: KUNDASANG: 1200 m, *RSNB 1389* (K, SING).

23.13.9. Microsorum sarawakense (Baker) Holttum, Rev. Fl. Malaya 2: 175, f. 85 (1955).

Polypodium sarawakense Baker, Bot. J. Linn. Soc. 22: 228 (1886). C. Chr. & Holttum, Gardens' Bull. 7: 307 (1934).

Epiphyte. Hill forest and lower montane forest. Elevation: 800–1400 m.

Material examined: DALLAS: 900 m, *Clemens 27349* (K), 1100 m, *Holttum SFN 25148* (K, SING); KIAU: *Topping 1574* (SING); KUNDASANG: 1200 m, *RSNB 1402* (K, SING); MAMUT RIVER: 1400 m, *RSNB 1720* (K, SING); MAMUT ROAD: 1200–1300 m, *Tamura & Hotta 358* (K); SAYAP: 800–1000 m, *Beaman 9806* (K, MICH, MSC); TAHUBANG RIVER: *Clemens 30703* (K); UPPER KINABALU: *Clemens 26942* (K).

23.13.10. Microsorum scolopendria (N. L. Burm.) Copel., Univ. Calif. Publ. Bot. 16: 112 (1929).

Lithophyte. Lowland dipterocarp forest. Elevation: 500 m.

Material examined: PORING HOT SPRINGS: 500 m, *Parris & Croxall 9003* (K).

23.13.11. Microsorum zippelii (Blume) Ching, Bull. Fan Mem. Inst. Biol. Bot. 4: 308 (1933).

Polypodium zippelii Blume, Fl. Javae 2: 172, t. 80 (1829). C. Chr. & Holttum, Gardens' Bull. 7: 307 (1934).

Epiphyte. Lower montane valley forest. Elevation: 900–1700 m.

Material examined: KUNDASANG: 1200 m, *RSNB 1420B* (SING), 1200 m, *1420b* (K); LIWAGU RIVER: 1400 m, *Parris & Croxall 8606* (K); MAMUT ROAD: 1200–1300 m, *Tamura & Hotta 371* (K); MINITINDUK: 900 m, *Holttum SFN 25589* (SING); PINOSUK PLATEAU: 1500 m, *Parris & Croxall 9147* (K); TAHUBANG RIVER: 1400 m, *Clemens 40296* (K); TENOMPOK: 1500 m, *Clemens 26929* (K), 1500 m, *29231* (K), 1400 m, *Holttum SFN 25297* (K, SING); WEST MESILAU RIVER: 1600–1700 m, *Beaman 8652* (K, MICH, MSC), 1600 m, *9359* (K, MICH, MSC).

23.14. PARAGRAMMA

Holttum, R. E. 1968. Revised Flora of Malaya. Vol. 2, Ferns. Ed. 2: 151–152 [as *Lepisorus*].

23.14.1. Paragramma longifolia (Blume) T. Moore, Index, 32 (1857).

Epiphyte and lithophyte. Lowland dipterocarp forest on ultramafics. Elevation: 500 m.

Material examined: PENATARAN RIVER: 500 m, *Beaman 8873* (K, MICH, MSC), 500 m, *9295* (K, MICH, MSC), 500 m, *9320* (K, MICH, MSC).

23.15. PHOTINOPTERIS

Roos, M. C. 1985. Phylogenetic systematics of the Drynarioideae (Polypodiaceae). Thesis, Leiden, University Printer; as *Aglaomorpha*.

23.15.1. Photinopteris speciosa (Blume) C. Presl, Epim., 264 (1849). C. Chr. & Holttum, Gardens' Bull. 7: 314 (1934).

Terrestrial ? and lithophytic. Lowland forest. Elevation: 300–900 m.

Material examined: DALLAS: 900 m, *Clemens 26005* (K); KAUNG: 300 m, *Clemens 26005* (L); PORING HOT SPRINGS: 500 m, *Parris & Croxall 8987* (K).

23.16. PHYMATOPTERIS

Holttum, R. E. 1968. Revised Flora of Malaya. Vol. 2, Ferns. Ed. 2: 193–197 [as *Crypsinus*].

23.16.1. Phymatopteris albidosquamata (Blume) Pichi Serm., Webbia 28: 461 (1973).

Polypodium albidosquamatum Blume, Enum. Pl. Javae, 132 (1828). C. Chr. & Holttum, Gardens' Bull. 7: 311 (1934).

Epiphyte. Lower montane forest. Elevation: 1100–1600 m.

Material examined: DALLAS/TENOMPOK: 1200 m, *Clemens 28153* (L, US); MENTEKI RIVER: 1600 m, *Beaman 10781* (MICH); PINOSUK PLATEAU: 1500 m, *Parris & Croxall 9143* (K); TAHUBANG RIVER: 1100 m, *Clemens 32388* (US); TENOMPOK: 1500 m, *Clemens 27904* (US), 1400 m, *Holttum SFN 25293* (SING).

23.16.2. Phymatopteris pakkaensis (C. Chr.) Parris, comb. nov.

Polypodium pakkaense C. Chr. in C. Chr. & Holttum, Gardens' Bull. 7: 310 (1934). Type: Paka-paka Cave: 3000 m, *Holttum SFN 25515* (holotype BM n.v.). *Polypodium taeniatum* Sw. var. *palmatum* (Blume) sensu C. Chr. & Holttum p.p. non Blume, Gardens' Bull. 7: 310 (1934).

Epiphyte and lithophyte. Tall upper montane forest and lower montane mossy forest. Elevation: 2100–3400 m. Endemic to Mount Kinabalu.

Material examined: EASTERN SHOULDER: 2900 m, *RSNB 805* (K), 3200 m, *879* (K), 2900 m, *Collenette 21555* (US); EASTERN SHOULDER, CAMP 4: 2700 m, *RSNB 1151* (K); GOKING'S VALLEY: 2800 m, *Fuchs 21485* (K); KEMBURONGOH: 2900 m, *Clemens 27962* (K, US), 2100 m, *Sinclair et al. 9055* (SING); KINATEKI RIVER HEAD: 2100 m, *Clemens 31871* (K); MESILAU BASIN: 2100–2400 m, *Clemens 28857* (K, US); MOUNT KINABALU: 3200 m, *Haviland 1490* (K); PAKA-PAKA CAVE: *Clemens 10587* (K), 2700 m, *Meijer SAN 22078* (K), 3000 m, *Sinclair et al. 9119* (K); SHANGRI LA VALLEY: 3400 m, *Collenette 21509* (K); SUMMIT TRAIL: 2500 m, *Parris 11549* (K), 3000 m, *Parris & Croxall 8727* (K), 2500 m, *8803* (K).

23.16.3. Phymatopteris taeniata (Sw.) Pichi Serm., Webbia 28: 465 (1973).

a. var. **taeniata**

Polypodium taeniatum Sw., Schrad. J. Bot. 1800 (2): 26 (1801). C. Chr. & Holttum, Gardens' Bull. 7: 309 (1934).

Epiphyte. Lower montane forest. Elevation: 1400–2700 m.

Material examined: EASTERN SHOULDER, CAMP 4: 2700 m, *RSNB 1151* (L); EASTERN SHOULDER, CAMP 3: 2400 m, *RSNB 1164* (L); GOLF COURSE SITE: 1800 m, *Beaman 7493* (MICH, MSC); KINATEKI RIVER HEAD: 2400 m, *Clemens 31787* (US); LIWAGU RIVER TRAIL: 1400 m, *Parris & Croxall 9047* (K); LUMU-LUMU: 2100 m, *Clemens 29971* (US); MEMPENING TRAIL: 1600 m, *Parris 11503* (K); SUMMIT TRAIL: 2000 m, *Parris & Croxall 8855* (K); TENOMPOK: 1500 m, *Beaman 10523* (MICH), 1500 m, *Clemens 28245* (L); WEST MESILAU RIVER: 1600 m, *Beaman 9033* (K, MICH, MSC).

b. var. **palmata** (Blume) Parris, comb. nov.

Polypodium palmatum Blume, Fl. Javae 2: 150 (1829). *P. taeniatum* var. *palmatum* (Blume) C. Chr. in C. Chr. & Holttum, Gardens' Bull. 7: 310 (1934).

Epiphyte? Three specimens cited by Christensen and Holttum have not been seen.

23.16.4. Phymatopteris triloba (Houtt.) Pichi Serm., Webbia 28: 465 (1973).

Polypodium incurvatum Blume, Enum. Pl. Javae, 126 (1828). C. Chr. & Holttum, Gardens' Bull. 7: 311 (1934). *Selliguea triloba* (Houtt.) M. G. Price, Contr. Univ. Mich. Herb. 17: 276 (1990).

Epiphyte and lithophyte. Lower montane forest. Elevation: 1200–2400 m.

Material examined: BAMBANGAN RIVER: 1500 m, *RSNB 4414* (SING), 1500 m, *RSNB 1305* (L, SING); DALLAS/TENOMPOK: 1200 m, *Clemens 27564* (US); EASTERN SHOULDER, CAMP 3: 2400 m, *RSNB 1165* (L, SING); KEMBURONGOH/LUMU-LUMU: 1800–2400 m, *Clemens 27933* (US); KIAU VIEW TRAIL: 1600 m, *Parris 10791* (K), 1600 m, *Parris & Croxall 9133* (K); MESILAU CAVE: 2100 m, *Collenette 21620* (L, US); PENIBUKAN: 1200–1800 m, *Clemens 40694* (L); PINOSUK PLATEAU: 1400 m, *Beaman 10732* (MICH), 1600 m, *de Vogel 8014* (L); SUMMIT TRAIL: 2100 m, *Parris & Croxall 8854* (K); TENOMPOK: 1400 m, *Holttum SFN 25418* (SING).

23.17. PLATYCERIUM

Hennipman, E. & M. C. Roos. 1982. A monograph of the fern genus *Platycerium* (Polypodiaceae). North-Holland Publishing Co., Amsterdam, Oxford, New York.

23.17.1. Platycerium coronarium (K. D. Koenig ex O. F. Muell.) Desv., Mem. Soc. Linn. Paris 6: 213 (1827). C. Chr. & Holttum, Gardens' Bull. 7: 314 (1934).

Epiphyte. Lowland dipterocarp forest. Elevation: 900 m.

Material examined: DALLAS: 900 m, *Clemens 26495* (K).

23.18. POLYPODIOPTERIS

Copeland, E. B. 1917. Keys to the ferns of Borneo. Sarawak Mus. J. 2: 398 [in *Polypodium* subgenus *Goniophlebium*]. Copeland, E. B. 1947. Genera Filicum. Ann. Crypto. Phytopath., Vol. 5: 210 [as *Polypodiopsis*].

23.18.1. Polypodiopteris brachypodia (Copel.) C. F. Reed, Amer. Fern J. 38: 87 (1948).

Polypodium brachypodium Copel., Philipp. J. Sci. 12C: 62 (1917). C. Chr. & Holttum, Gardens' Bull. 7: 305 (1934). Type: Gurulau Spur: *Topping 1823* (holotype PNH?†; isotype MICH!).

Epiphyte. Lower montane forest. Elevation: 1500–2300 m. Not endemic to Mount Kinabalu.

Additional material examined: KIAU VIEW TRAIL: 1600 m, *Parris 10811* (K), 1600 m, *Parris & Croxall 9128* (K); KILEMBUN BASIN: 2300 m, *Clemens 33692* (K, SING); KILEMBUN RIVER HEAD: 1800 m, *Clemens 32467* (K, SING); PARK HEADQUARTERS: 1500 m, *Parris & Croxall 8474* (K); TENOMPOK: 1500 m, *Beaman 10524* (MICH), 1500 m, *Clemens 28344* (SING), 1500 m, *28544* (K).

23.19. PYCNOLOMA

Christensen, C. 1929. Taxonomic fern-studies I. Revision of the polypodioid genera with longitudinal coenosori (Cochlidiinae and "Drymoglossinae"); with a discussion of their phylogeny. Dansk Bot. Arkiv 6: 75–80.

23.19.1. Pycnoloma metacoelum (Alderw.) C. Chr., Dansk Bot. Arkiv 6: 77, t. 8–10 (1929).

Epiphyte. Montane dipterocarp forest. Elevation: 1100 m.

Material examined: LOHAN/MAMUT COPPER MINE: 1100 m, *Beaman 10658* (MICH, MSC).

23.19.2. Pycnoloma murudense C. Chr., Dansk Bot. Arkiv. 6: 78, t. 8, 10 (1929). C. Chr. & Holttum, Gardens' Bull. 7: 311 (1934).

Epiphyte. Lower montane forest. Elevation: 1200–1600 m.

Material examined: KILEMBUN RIVER: 1500 m, *Clemens 33980* (K); MAMUT COPPER MINE: 1400 m, *Collenette 1029* (K); PARK HEADQUARTERS: 1400 m, *Edwards 2184* (K), 1500 m, *Parris & Croxall 8573* (K); PENIBUKAN: 1200–1500 m, *Clemens 31365* (SING); PINOSUK PLATEAU: 1600 m, *RSNB 1786* (K, SING); TAHUBANG RIVER: *Haviland 1474* (K).

23.20. PYRROSIA

Hovenkamp, P. 1986. A monograph of the fern genus *Pyrrosia*. E. J. Brill/Leiden Univ. Press (Leiden Bot. Ser., Vol. 9).

23.20.1. Pyrrosia christii (Giesenh.) Ching, Bull. Chin. Bot. Soc. 1: 58 (1935).

Cyclophorus christii (Giesenh.) C. Chr., Index Fil., 198 (1905). C. Chr. & Holttum, Gardens' Bull. 7: 313 (1934).

Epiphyte. Montane dipterocarp forest. Elevation: 800–1200 m.

Material examined: DALLAS: 800 m, *Holttum s.n.* (SING); MAMUT RIVER: 1200 m, *RSNB 1727* (K, SING).

23.20.2. Pyrrosia kinabaluensis Hovenkamp, Blumea 30: 208 (1984). Type: Tenompok: 1500 m, *Clemens 26984* (holotype L n.v.; isotypes K!, SING!).

Cyclophorus borneensis sensu C. Chr. & Holttum p.p. non Copel., Gardens' Bull. 7: 313 (1934).

Epiphyte. Lower montane valley forest. Elevation: 1500 m. Not endemic to Mount Kinabalu.

Additional material examined: PARK HEADQUARTERS: 1500 m, *Parris 10780* (K); TENOMPOK: 1500 m, *Clemens 27479* (K), 1500 m, *27479b* (K).

23.20.3. Pyrrosia lanceolata (L.) Farw., Amer. Midl. Nat. 12: 245 (1930).

Cyclophorus adnascens (Sw.) Desv., Mag. Ges. Naturf. Freunde Berlin 5: 300 (1811). C. Chr. & Holttum, Gardens' Bull. 7: 313 (1934). *Cyclophorus varius* (Kaulf.)

Gaud. in Freyc., Voy. Uranie, 364 (1829). C. Chr. & Holttum, Gardens' Bull. 7: 313 (1934).

Epiphyte. Lowland dipterocarp forest to lower montane forest. Elevation: 500–1500 m.

Material examined: DALLAS: *Clemens 27293* (US), 900 m, *29299* (US); KAUNG: 500 m, *Holttum SFN 25123* (K, SING); KUNDASANG: 1200 m, *RSNB 1404* (K, SING); LOHAN RIVER: 800–1000 m, *Beaman 9058* (K, MICH, MSC); LOHAN/MAMUT COPPER MINE: 900 m, *Beaman 10624* (K, MICH, MSC); PINOSUK PLATEAU: 1500 m, *Parris & Croxall 9175* (K); PORING HOT SPRINGS: 500 m, *Parris & Croxall 8985* (K); PORING HOT SPRINGS/LANGANAN WATER FALLS: 700 m, *Parris & Croxall 8945* (K); TENOMPOK: 1500 m, *Clemens 28685* (K), 1500 m, *29628* (K).

23.20.4. Pyrrosia platyphylla Hovenkamp, Blumea 30: 207 (1984).

Epiphyte. Lowland dipterocarp forest. Elevation: 500–900 m.

Material examined: LOHAN RIVER: 800 m, *Beaman 8364* (K, MICH, MSC), 700–900 m, *9243* (K, MICH, MSC); PORING HOT SPRINGS: 500 m, *Parris & Croxall 8983* (K).

23.20.5. Pyrrosia rasamalae (Racib.) Shing, Amer. Fern J. 73: 78 (1983).

Cyclophorus borneensis Copel., Philipp. J. Sci. 12C: 64 (1917). C. Chr. & Holttum, Gardens' Bull. 7: 313 (1934) p.p. Type: Kiau: *Topping 1508* (holotype PNH?†; isotype GH n.v.). *Cyclophorus flocciger* sensu C. Chr. & Holttum non (Blume) C. Presl, Gardens' Bull. 7: 313 (1934).

Epiphyte. Montane dipterocarp forest. Elevation: 800–1500 m.

Material examined: LOHAN/MAMUT COPPER MINE: 900 m, *Beaman 10625* (K, MICH, MSC); SAYAP: 800–1000 m, *Beaman 9807* (MICH, MSC); TAHUBANG RIVER: 900 m, *Clemens 33229* (K); TENOMPOK: 1500 m, *Clemens 27327* (K), 1400 m, *Holttum SFN 25358* (K, SING).

23.21. SELLIGUEA

Holttum, R. E. 1968. Revised Flora of Malaya. Vol. 2, Ferns. Ed. 2: 156–159.

23.21.1. Selliguea heterocarpa (Blume) Blume, Enum. Pl. Javae addenda, 2 (1828).

Epiphyte. Lowland forest. Elevation: 600 m.

Material examined: NALUMAD: 600 m, *Shea & Aban SAN 77280* (K).

23.21.2. Selliguea kamborangana (C. Chr.) M. G. Price, Contr. Univ. Mich. Herb. 16: 198 (1987).

Polypodium kamboranganum C. Chr. in C. Chr. & Holttum, Gardens' Bull. 7: 306 (1934). Type: Kemburongoh: *Holttum SFN 25543* (holotype BM n.v.; isotype SING!).

Epiphyte ? Lower montane forest. Elevation: 2100 m. Known also from Palawan fide Price.

23.22. THYLACOPTERIS

Holttum, R. E. 1968. Revised Flora of Malaya. Vol. 2, Ferns. Ed. 2: 203–204 [as *Polypodium*]].

23.22.1. Thylacopteris papillosa (Blume) Kunze ex J. Sm., Hist. Fil., 88 (1875).

Polypodium papillosum Blume, Enum. Pl. Javae, 131 (1828). C. Chr. & Holttum, Gardens' Bull. 7: 304 (1934).

Epiphyte. Lowland dipterocarp forest to lower montane forest. Elevation: 500–1800 m.

Material examined: DALLAS: 900 m, *Clemens 26892* (K), 900 m, *27290* (K), 1100 m, *Holttum SFN 25143* (K, SING); EASTERN SHOULDER: 1100 m, *RSNB 636* (K, SING); KIAU/LUBANG: *Topping 1588* (SING, US); KILEMBUN BASIN: 1400 m, *Clemens 34392* (K); KILEMBUN RIVER: 1500–1800 m, *Clemens 34392* (SING); KIPUNGIT FALLS: 500 m, *Parris & Croxall 8912* (K); LIWAGU/MESILAU RIVERS: 1200 m, *RSNB 2859* (K, SING); TENOMPOK: 1500 m, *Clemens 26488* (K), 1500 m, *27290b* (K).

24. PTERIDACEAE

24.1. PTERIS

Copeland, E. B. 1917. Keys to the ferns of Borneo. Sarawak Mus. J. 2: 332–335. Holttum, R. E. 1968. Revised Flora of Malaya. Vol. 2, Ferns. Ed. 2: 393–409. Tagawa, M. & Iwatsuki, K. 1985. Flora of Thailand. Vol. 3 (2): 231–257.

24.1.1. Pteris biaurita L., Sp. Pl. 1076 (1753). C. Chr. & Holttum, Gardens' Bull. 7: 286 (1934).

Terrestrial. Elevation: 900 m.

Material examined: DALLAS: 900 m, *Clemens 27068* (K), 900 m, *27213* (K); KAUNG/KIAU: *Topping 1505* (US); KIAU: 900 m, *Holttum SFN 25595* (K).

24.1.2. Pteris clemensiae Copel., Philipp. J. Sci. 12C: 47 (1917). C. Chr. & Holttum, Gardens' Bull. 7: 285 (1934). Type: Lubang: *Clemens 10348* (holotype PNH?†; isotypes MICH!, UC!).

Terrestrial. Lower montane forest. Elevation: 1200–1700 m. Endemic to Mount Kinabalu.

Additional material examined: KIAU/LUBANG: *Topping 1602* (MICH, US); KILEMBUN RIVER HEAD: 1400 m, *Clemens 32514* (MICH); LUBANG: 1200 m, *Holttum SFN 25722* (SING); TENOMPOK: 1500 m, *Clemens 27909* (K, SING), 1700 m, *28095* (K), 1600 m, *29311* (K, SINC).

24.1.3. Pteris digitata (Baker) C. Chr., Dansk Bot. Arkiv 9: 71 (1937).

Pteris quadriaurita Retz. var. *digitata* Baker, J. Bot. 17: 40 (1879). *Pteris grevilleana* sensu C. Chr. & Holttum non Wall. ex J. G. Agardh, Gardens' Bull. 7: 285 (1934).

Terrestrial? Lower montane forest? Elevation: 1200–1500 m.

Material examined: KIAU/LUBANG: *Topping 1605* (US); LUBANG: 1200 m, *Holttum SFN 25556* (K, SING); PENIBUKAN: 1200 m, *Clemens 30622* (K, SING); TENOMPOK: 1500 m, *Clemens 28172* (US).

24.1.4. Pteris ensiformis N. L. Burm., Fl. Ind., 230 (1768). C. Chr. & Holttum, Gardens' Bull. 7: 285 (1934).

Terrestrial. Hill forest. Elevation: 300–1500 m.

Material examined: DALLAS: 900 m, *Clemens 27485* (K); KEBAYAU/KAUNG: 300 m, *Clemens 27690* (K); KIAU: *Topping 1558* (US); PORING HOT SPRINGS/LANGANAN WATER FALLS: 700 m, *Parris & Croxall 8944* (K); TENOMPOK: 1500 m, *Clemens 29028* (K).

24.1.5. Pteris excelsa Gaud. in Freyc., Voy. Bot., 388 (1827). C. Chr. & Holttum, Gardens' Bull. 7: 286 (1934).

Terrestrial. Hill forest. Elevation: 800–1400 m.

Material examined: DALLAS: 800 m, *Holttum SFN 25275* (K, SING); KIAU: *Topping 1568* (SING, US); KINATEKI RIVER: 1200 m, *Clemens 31086* (K); LIWAGU RIVER TRAIL: 1400 m, *Parris & Croxall 9028* (K); TENOMPOK: 1200 m, *Clemens 27118* (US).

24.1.6. Pteris flava Goldm., Nova Acta 19, Suppl. 1: 457 (1843). C. Chr. & Holttum, Gardens' Bull. 7: 286 (1934).

Terrestrial. Lower montane forest. Elevation: 1400–2500 m.

Material examined: KADAMAIAN FALLS TRAIL: 1700 m, *Parris & Croxall 9096* (K); KEMBURONGOH: 2100 m, *Holttum SFN 25521* (K, SING); KILEMBUN RIVER HEAD: 1400 m, *Clemens 32444* (K); SUMMIT TRAIL: 2500 m, *Parris 11609* (K), 2400 m, *Parris & Croxall 8860* (K).

24.1.7. Pteris holttumii C. Chr. in C. Chr. & Holttum, Gardens' Bull. 7: 287 (1934). Type: Dallas: 800 m, *Holttum SFN 25363* (holotype BM n.v.; isotypes CGE!, K!, SING!).

Terrestrial. Hill forest. Elevation: 600–800 m. Not endemic to Mount Kinabalu.

Additional material examined: KAUNG: 600 m, *Clemens 27229* (K).

24.1.8. Pteris kinabaluensis C. Chr. in C. Chr. & Holttum, Gardens' Bull. 7: 286 (1934). Type: Marai Parai Spur: 1800 m, *Holttum SFN 25614* (holotype BM n.v.; isotypes K!, SING!).

Terrestrial. Lower montane forest. Elevation: 1700–2100 m. Not endemic to Mount Kinabalu.

Additional material examined: KADAMAIAN FALLS TRAIL: 1700 m, *Parris & Croxall 9095* (K); LUMU-LUMU: 2100 m, *Clemens 29973* (K, SING); TENOMPOK: 2100 m, *Clemens 29567* (K).

24.1.9. Pteris longipes D. Don, Prodr. Fl. Nepal, 15 (1825).

Pteris longipes D. Don var. *philippinensis* (Fée) C. Chr. in C. Chr. & Holttum, Gardens' Bull. 7: 286 (1934).

Terrestrial. Hill forest. Elevation: 800–1500 m.

Material examined: DALLAS: 800 m, *Clemens 27289* (K, SING); KADAMAIAN RIVER NEAR MINITINDUK: 900 m, *Holttum SFN 25569* (K); KILEMBUN RIVER: 1200 m, *Clemens 34013* (K); MARAI PARAI: 1500 m, *Clemens 32412* (K); MINITINDUK: 900 m, *Holttum SFN 25569* (SING); TAKUTAN: 800 m, *Shea & Aban SAN 77177* (K).

24.1.10. Pteris longipinnula Wall. ex J. G. Agardh, Recen., 19 (1839). C. Chr. & Holttum, Gardens' Bull. 7: 285 (1934).

Terrestrial. Lower montane forest. Elevation: 2700 m.

Material examined: EASTERN SHOULDER: 2700 m, *RSNB 950* (K, SING).

24.1.11. Pteris mertensioides Willd., Sp. Pl. 5: 394 (1810).

Pteris decussata J. Sm., J. Bot. 3: 405 (1841) nom. nud. C. Chr. & Holttum, Gardens' Bull. 7: 286 (1934).

Terrestrial. Lowland dipterocarp forest to lower montane forest. Elevation: 600–1700 m.

Material examined: KADAMAIAN FALLS TRAIL: 1700 m, *Parris & Croxall 9083* (K); KIAU/LUBANG: *Topping 1613* (SING, US); KIPUNGIT FALLS: 600 m, *Parris & Croxall 8903* (K); LUBANG: 1200 m, *Holttum SFN 25558* (K, SING); MOUNT KINABALU: 1200 m, *Low s.n.* (K); PENIBUKAN: 1200 m, *Clemens 40307* (K).

24.1.12. Pteris purpureorachis Copel., Philipp. J. Sci. 12C: 48 (1917). C. Chr. & Holttum, Gardens' Bull. 7: 286 (1934). Type: Lubang: *Clemens 10350* (holotype PNH?†; isotype MICH!).

Terrestrial. Hill forest to lower montane forest. Elevation: 900–1700 m. Not endemic to Mount Kinabalu.

Additional material examined: DALLAS: 1200 m, *Clemens 27730* (K, MICH); KIAU VIEW TRAIL: 1600 m, *Parris 10793* (K); LIWAGU/MESILAU RIVERS: 1400 m, *RSNB 2765* (K); PENIBUKAN: 1200 m, *Clemens 51710* (K); TENOMPOK: 1700 m, *Clemens 28093* (K), 1400 m, *Holttum SFN 25413* (K); ULAR HILL TRAIL: 1700 m, *Parris 11449* (K).

24.1.13. Pteris rangiferina C. Presl ex Miq., Ann. Lugd. Bat. 4: 95 (1868–69).

Pteris rangiferina C. Presl ex Miq. var. *scabripes* C. Chr. in C. Chr. & Holttum, Gardens' Bull. 7: 285 (1934). Type: Marai Parai Spur: *Clemens 11062* (syntype BM n.v.); Penibukan: 1200 m, *Holttum SFN 25621* (syntype BM n.v.; isosyntypes K!, SING!).

Terrestrial. Lower montane forest, apparently mostly on ultramafics. Elevation: 800–2100 m.

Additional material examined: MARAI PARAI SPUR: 1500 m, *Clemens 32403* (US); MESILAU CAVE: 2000–2100 m, *Beaman 8144* (K, MICH, MSC), 2000–2100 m, *9130* (K, MICH, MSC), 2100 m, *Collenette 21611* (US); MT. NUNGKEK: 800 m, *Clemens 32738* (K); PENIBUKAN: 1200 m, *Parris 11519* (K).

24.1.14. Pteris tripartita Sw., Schrad. J. Bot. 1800 (2): 67 (1801).

Terrestrial. Lowland dipterocarp forest and lower montane forest. Elevation: 500–1500 m.

Material examined: PENATARAN BASIN: 1200 m, *Clemens 34142* (K); PENATARAN RIVER: 1500 m, *Clemens 32562* (K); PINOSUK PLATEAU: 1500 m, *Parris & Croxall 9156* (K); PORING HOT SPRINGS: 500 m, *Parris & Croxall 9010* (K).

24.1.15. Pteris vittata L., Sp. Pl. 1074 (1753). C. Chr. & Holttum, Gardens' Bull. 7: 285 (1934).

Terrestrial. Lower montane forest.

Material examined: LUBANG: *Clemens 10323* (BM).

24.1.16. Pteris wallichiana J. G. Agardh, Recen., 69 (1839).

Terrestrial. Lower montane forest. Elevation: 2100–2900 m.

Material examined: EASTERN SHOULDER: 2900 m, *RSNB 946* (K, SING); MESILAU RIVER: 2100 m, *Collenette 21651* (K).

24.1.17. Pteris sp. 1

Terrestrial. Lowland dipterocarp forest. Elevation: 600 m. Endemic to Mount Kinabalu.

Material examined: PORING HOT SPRINGS/LANGANAN WATER FALLS: 600 m, *Parris & Croxall 8947* (K).

24.1.18. Pteris sp. 2

Terrestrial. Lower montane forest. Elevation: 1700 m. Endemic to Mount Kinabalu.

Material examined: KADAMAIAN FALLS TRAIL: 1700 m, *Parris & Croxall 9097* (K).

25. SCHIZAEACEAE

Holttum, R. E. 1959. Schizaeaceae. Fl. Males. 2, 1 (1): 37–61.

25.1. LYGODIUM

25.1.1. Lygodium circinnatum (N. L. Burm.) Sw., Syn. Fil., 153 (1806). C. Chr. & Holttum, Gardens' Bull. 7: 210 (1934).

Terrestrial, climber. Lowland forest. Elevation: 500 m.

Material examined: KAUNG: *Topping 1894* (MICH); KIAU/LUBANG: *Topping 1800* (MICH); TAKUTAN: 500 m, *Shea & Aban SAN 77185* (K).

25.1.2. Lygodium microphyllum (Cav.) R. Br., Prodr., 162 (1810).

Lygodium scandens Sw., Schrad. J. Bot. 1800 (2): 106 (1801). C. Chr. & Holttum, Gardens' Bull. 7: 210 (1934).

Terrestrial, climber. *Topping 1509* (BM n.v.), cited by Christensen and Holttum.

25.2. SCHIZAEA

25.2.1. Schizaea dichotoma (L.) J. Sm., Mem. Ac. Turin 5: 422, t. 9 (1793). C. Chr. & Holttum, Gardens' Bull. 7: 210 (1934).

Terrestrial. Hill forest on ultramafics. Elevation: 800–1000 m.

Material examined: LOHAN RIVER: 800–1000 m, *Beaman 9067* (MICH); MELANGKAP TOMIS: 900–1000 m, *Beaman 8987* (MICH).

25.2.2. Schizaea digitata (L.) Sw., Syn. Fil., 150, 380, t. 4, f. 1 (1806). C. Chr. & Holttum, Gardens' Bull. 7: 210 (1934).

Terrestrial. Hill forest. Elevation: 900–1400 m.

Material examined: DALLAS: 900 m, *Clemens 27357* (K); LIWAGU RIVER TRAIL: 1400 m, *Edwards 2205* (K).

25.2.3. Schizaea fistulosa Labill., Nov. Holl. Pl. Spec. 2: 1103, t. 250, f. 3 (1806). C. Chr. & Holttum, Gardens' Bull. 7: 210 (1934).

Terrestrial. Low upper montane forest on ultramafics. Elevation: 1500–3000 m.

Material examined: KEMBURONGOH/PAKA-PAKA CAVE: 2600 m, *Sinclair et al. 9100* (K, SING); LUBANG/PAKA-PAKA CAVE: *Clemens 10729* (K); MOUNT KINABALU: 1500 m, *Burbidge s.n.* (K); PAKA-PAKA CAVE: 2900 m, *Clemens 27952* (K), 2700 m, *Holttum SFN 25505* (K), *Topping 1720* (SING), *1724* (SING); PIG HILL: 2000–2300 m, *Beaman 9885* (K, MICH, MSC); SUMMIT TRAIL: 2500–2900 m, *Jacobs 5732* (K), 3000 m, *Parris 11569* (K), 2900 m, *Parris & Croxall 8730* (K).

25.2.4. Schizaea malaccana Baker

a. var. **robustior** C. Chr. in C. Chr. & Holttum, Gardens' Bull. 7: 210 (1934). Type: Marai Parai Spur: *Clemens 10919* (holotype BM? n.v.; isotype K!).

Terrestrial and epiphytic. Lower montane forest. Elevation: 1500 m. Not endemic to Mount Kinabalu.

Additional material examined: MARAI PARAI: 1500 m, *Clemens 51761* (K).

26. THELYPTERIDACEAE

Holttum, R. E. 1981. Thelypteridaceae. Fl. Males. 2, 1 (5): 331–599.

26.1. AMPHINEURON

26.1.1. Amphineuron immersum (Blume) Holttum in Nayar & Kaur, Comp. to Bedd., 203 (1974).

Dryopteris immersa (Blume) Kuntze, Rev. Gen. Pl. 2: 813 (1891). C. Chr. & Holttum, Gardens' Bull. 7: 243 (1934).

Terrestrial. Lowland dipterocarp and hill forest, possibly also lower montane forest. Elevation: 500–1500 m.

Material examined: DALLAS: 900 m, *Clemens 27739* (K); DALLAS/TENOMPOK: 1400 m, *Clemens 27523* (US); KEBAYAU/KAUNG: *Topping 1899* (SING, US); KUNDASANG: 1200 m, *Clemens 29765* (K); PORING HOT SPRINGS: 500 m, *Parris & Croxall 9006* (K); TENOMPOK: 1500 m, *Clemens 29523* (K).

26.1.2. Amphineuron kiauense (C. Chr.) Holttum, Fl. Males. 2, 1: 550 (1981).

Terrestrial. Lower montane forest. Elevation: 1500 m.

Material examined: TENOMPOK: 1500 m, *Clemens 29765a* (K).

26.2. CHINGIA

26.2.1. Chingia atrospinosa (C. Chr.) Holttum, Kalikasan 3: 19 (1974).

Dryopteris ferox sensu C. Chr. & Holttum non (Blume) Kuntze, Gardens' Bull. 7: 246 (1934).

Terrestrial. Hill forest and lower montane forest. Elevation: 1100–2100 m.

Material examined: KILEMBUN BASIN: 1100 m, *Clemens 34105* (K); MAMUT RIVER: 1200 m, *RSNB 1216* (K, SING), 1200 m, *1679* (K, SING); MESILAU BASIN: 2100 m, *Clemens 29062* (K, SING); PENATARAN BASIN: 1500 m, *Clemens 32564* (SING), 1100 m, *34105* (SING); ULAR HILL TRAIL: 1800 m, *Parris & Croxall 9087* (K); WEST MESILAU RIVER: 1600–1800 m, *Kokawa & Hotta 4329* (K).

26.2.2. Chingia clavipilosa Holttum, Kalikasan 3: 23 (1974).

a. var. **clavipilosa** Type: Kemburongoh: 2300 m, *Holttum 44* (holotype K!; isotype SING!).

Terrestrial. Lower montane forest. Elevation: 1900–2300 m. Not endemic to Mount Kinabalu.

Additional material examined: MOUNT KINABALU: 1900 m, *Holttum 52* (K); SUMMIT TRAIL: 2300 m, *Parris & Croxall 8863* (K).

26.3. CHRISTELLA

26.3.1. Christella arida (D. Don) Holttum in Nayar & Kaur, Comp. to Bedd., 206 (1974).

Dryopteris arida (D. Don) Kuntze, Rev. Gen. Pl. 2: 812 (1891). C. Chr. & Holttum, Gardens' Bull. 7: 245 (1934).

Terrestrial. Hill forest?

Material examined: KIAU: *Topping 1560* (MICH).

26.3.2. Christella hispidula (Decne.) Holttum, Kew Bull. 31: 312 (1976).

Dryopteris contigua Rosenstock, Meded. Rijksherb. 31: 7 (1917). C. Chr. & Holttum, Gardens' Bull. 7: 244 (1934).

Terrestrial. Lowland dipterocarp forest to lower montane forest. Elevation: 600–1500 m.

Material examined: DALLAS: 900 m, *Clemens 27475* (K), 900 m, *27739* (US); KUNDASANG: 900 m, *Clemens 29238* (K); MOUNT KINABALU: *Holttum SFN 25727* (K, SING); PARK HEADQUARTERS: 1500 m, *Parris 10832* (K); PORING HOT SPRINGS/LANGANAN WATER FALLS: 700 m, *Parris & Croxall 8921* (K), 600 m, *8953* (K); TENOMPOK: 1500 m, *Clemens 27469* (K).

26.3.3. Christella parasitica (L.) Lév., Fl. Kouytcheou, 475 (1915).

Dryopteris parasitica (L.) Kuntze, Rev. Gen. Pl. 2: 811 (1891). C. Chr. & Holttum, Gardens' Bull. 7: 244 (1934).

Terrestrial. Lowland and hill forest. Elevation: 500–900 m.

Material examined: KADAMAIAN RIVER NEAR MINITINDUK: 900 m, *Holttum SFN 25574* (K, SING); KIAU: *Topping 1513* (US); TAKUTAN: 500 m, *Shea & Aban SAN 77191* (K).

26.3.4. Christella subpubescens (Blume) Holttum, Webbia 30: 193 (1976).

Dryopteris subpubescens (Blume) C. Chr., Gardens' Bull. 4: 390 (1929). C. Chr. & Holttum, Gardens' Bull. 7: 244 (1934).

Terrestrial. Hill forest. Elevation: 300–900 m.

Material examined: KADAMAIAN RIVER NEAR MINITINDUK: 900 m, *Holttum SFN 25575* (K, SING); KAUNG: 300 m, *Holttum SFN 25121* (SING).

26.4. CORYPHOPTERIS

26.4.1. Coryphopteris badia (Alderw.) Holttum, Blumea 23: 44 (1976).

Dryopteris badia Alderw., Bull. Jard. Bot. Buit. 2, 16: 9 (1914). C. Chr. & Holttum, Gardens' Bull. 7: 241 (1934). *Dryopteris linearis* Copel., Philipp. J. Sci. 12C: 56 (1917). Type: Marai Parai Spur: *Clemens 11069* (lectotype of Holttum, Blumea 23: 44 (1976), MICH!; isolectotype K!).

Terrestrial. Lower montane forest, apparently on ultramafics. Elevation: 2100 m.

Additional material examined: KINATEKI RIVER HEAD: 2100 m, *Clemens 31770* (K).

26.4.2. Coryphopteris gymnopoda (Baker) Holttum, Blumea 23: 29 (1976).

a. var. **gymnopoda**

Nephrodium gymnopodum Baker in Stapf, Trans. Linn. Soc. Bot. 4: 252 (1894). Type: Mount Kinabalu: 3200 m, *Haviland 1486* (holotype K!). *Dryopteris kinabaluensis* Copel., Philipp. J. Sci. 12C: 55 (1917). Type: Paka-paka Cave: *Topping 1719* (lectotype of Holttum, Blumea 23: 29 (1976), MICH!). *Dryopteris viscosa* sensu C. Chr. & Holttum non (J. Sm.) Kuntze, Gardens' Bull. 7: 240 (1934). *Dryopteris viscosa* (J. Sm.) Kuntze var. *kamborangana* C. Chr. in C. Chr. & Holttum, Gardens' Bull. 7: 240 (1934). Type: Kemburongoh/Lumu-lumu: 1800 m, *Holttum SFN 25472* (syntype BM n.v.; isosyntypes K!, SING!); Marai Parai Spur: 1800 m, *Holttum SFN 25609* (syntype BM n.v.; isosyntype K!).

Terrestrial. Lower and upper montane forest, frequently on ultramafics. Elevation: 1400–3400 m. Not endemic to Mount Kinabalu.

Additional material examined: DACHANG: 3000 m, *Clemens 29066* (K); EASTERN SHOULDER: 3000 m, *RSNB 730* (K, SING); EASTERN SHOULDER, CAMP 4: 2700 m, *RSNB 1143* (K, SING), 2900 m, *1154* (K, SING); GURULAU SPUR: 3400 m, *Clemens 50896* (K); KEMBURONGOH: 2400 m, *Clemens 28964* (US), 2100 m, *Sinclair et al. 9047* (K, SING); KIAU VIEW TRAIL: 1600 m, *Parris 10804* (K); KILEMBUN BASIN: 1800 m, *Clemens 33675* (SING); LIWAGU RIVER HEAD: 2300 m, *Meijer SAN 24138* (K); MAMUT COPPER MINE: 1600–1700 m, *Beaman 9947* (MICH, MSC); MAMUT HILL: 1400–1600 m, *Kokawa & Hotta 5435* (K); MARAI PARAI SPUR: *Clemens 11068* (K); MESILAU CAMP: 1500 m, *Holttum 13* (K, SING); PAKA-PAKA CAVE: 3000 m, *Clemens 27971* (US), 3000 m, *Holttum SFN 25512* (K, SING); PAKA-PAKA CAVE/PANAR LABAN: *Kokawa & Hotta 3455* (K); PARK HEADQUARTERS: 1500 m, *Parris 10840* (K); PENATARAN BASIN: 1500–1800 m, *Clemens 33683* (K); SUMMIT TRAIL: 2900 m, *Collenette 21556* (US), 2200–2500 m, *Jacobs 5770* (US), 2300 m, *Parris 11539* (K), 3000 m, *Parris & Croxall 8729* (K), 3000 m, *8793* (K); ULAR HILL TRAIL: 1700 m, *Parris 11445* (K); UPPER KINABALU: *Clemens 29333* (K).

26.4.3. Coryphopteris multisora (C. Chr.) Holttum, Blumea 23: 26 (1976).

Dryopteris multisora C. Chr. in C. Chr. & Holttum, Gardens' Bull. 7: 241 (1934). Type: Kemburongoh: 2100 m, *Holttum SFN 25523* (holotype BM n.v.; isotypes K!, SING!).

Terrestrial. Lower montane forest, one record apparently from upper montane forest. Elevation: 1400–3400 m. Not endemic to Mount Kinabalu.

Additional material examined: KIAU VIEW TRAIL: 1600 m, *Parris 10803* (K); MESILAU CAMP: 1500 m, *Holttum 5* (K, SING); PAKA-PAKA CAVE: 3400 m, *Clemens 27970* (K); PARK HEADQUARTERS: 1600 m, *Parris & Croxall 8537* (K); SOSOPODON: 1400–1500 m, *Kokawa & Hotta 4535* (K); TENOMPOK: 1700 m, *Clemens 28098* (K), 1400 m, *Holttum SFN 25417* (K, SING).

26.4.4. Coryphopteris obtusata (Alderw.) Holttum, Blumea 23: 30 (1976).

Dryopteris supravillosa C. Chr. in C. Chr. & Holttum, Gardens' Bull. 7: 241 (1934). Type: Kemburongoh/Lumu-lumu: 1800 m, *Holttum 25471* (holotype BM n.v.; isotypes K!, SING!).

Terrestrial. Lower montane forest. Elevation: 1600–1800 m.

Additional material examined: KIAU VIEW TRAIL: 1600 m, *Parris 10801* (K), 1600 m, *11619* (K).

26.5. MACROTHELYPTERIS

26.5.1. Macrothelypteris multiseta (Baker) Ching, Acta Phytotax. Sinica 8: 309 (1963).

Dryopteris multiseta (Baker) C. Chr., Index Fil., 279 (1905). C. Chr. & Holttum, Gardens' Bull. 7: 243 (1934).

Terrestrial. Hill forest. Elevation: 1200 m.

Material examined: DALLAS/TENOMPOK: 1200 m, *Holttum SFN 25294* (K, SING).

26.5.2. Macrothelypteris torresiana (Gaud.) Ching, Acta Phytotax. Sinica 8: 310 (1963).

Dryopteris setigera sensu C. Chr. & Holttum non (Blume) Kuntze, Gardens' Bull. 7: 243 (1934).

Terrestrial. Hill forest to lower montane forest. Elevation: 800–1700 m.

Material examined: DALLAS: 800 m, *Holttum SFN 25369* (K, SING); KUNDASANG: 900 m, *Clemens 29242* (K); LUBANG: *Topping 1762* (SING, US); TENOMPOK: 1500–1700 m, *Kokawa & Hotta 3035* (US).

26.6. MESOPHLEBION

26.6.1. Mesophlebion crassifolium (Blume) Holttum, Blumea 22: 232 (1975).

Dryopteris crassifolia (Blume) Kuntze, Rev. Gen. Pl. 2: 812 (1891). C. Chr. & Holttum, Gardens' Bull. 7: 242 (1934). *Dryopteris crassifolia* (Blume) Kuntze var. *purpureolilacina* C. Chr. in C. Chr. & Holttum, Gardens' Bull. 7: 242 (1934). Type: Penibukan: 1200 m, *Holttum SFN 25599* (holotype BM n.v.; isotype SING!).

Terrestrial. Hill forest to lower montane forest. Elevation: 900–2400 m.

Additional material examined: DALLAS: 900 m, *Clemens 27526* (K), 1100 m, *Holttum SFN 25254* (SING); LIWAGU RIVER TRAIL: 1500 m, *Parris 10842* (K); PENATARAN RIVER: 2400 m, *Clemens 32546* (SING); TENOMPOK/LUMU-LUMU: 1500 m, *Holttum SFN 25440* (K, SING).

26.6.2. Mesophlebion dulitense Holttum, Blumea 22: 229 (1975).

Terrestrial. Lower montane forest. Elevation: 1500 m.

Material examined: KILEMBUN BASIN: *Clemens 33695* (K, SING); MEMPENING TRAIL: 1500 m, *Parris & Croxall 9119* (K); MESILAU CAMP: 1500 m, *Holttum 20* (K); MOUNT KINABALU: 1500 m, *Holttum 20* (SING).

26.7. METATHELYPTERIS

26.7.1. Metathelypteris dayii (Baker) Holttum in Nayar & Kaur, Comp. to Bedd., 205 (1974).

Terrestrial. Lower montane forest. Elevation: 1500–1600 m.

Material examined: KIAU VIEW TRAIL: 1600 m, *Parris 10795* (K), 1600 m, *10805* (K), 1600 m, *11617* (K); MEMPENING TRAIL: 1600 m, *Parris 11508* (K); PARK HEADQUARTERS: 1500 m, *Parris 11425* (K), 1500 m, *Parris & Croxall 8533* (K), 1500 m, *8535* (K), 1600 m, *9125* (K).

26.7.2. Metathelypteris flaccida (Blume) Ching, Acta Phytotax. Sinica 8: 306 (1963).

Terrestrial. Lower montane forest. Elevation: 1500–2000 m.

Material examined: KEMBURONGOH: 2000 m, *Molesworth-Allen 3235* (K); LIWAGU RIVER TRAIL: 1700 m, *Parris & Croxall 9106* (K); PARK HEADQUARTERS: 1500 m, *Parris 10831* (K), 1500 m, *Parris & Croxall 8561* (K); ULAR HILL TRAIL: 1700 m, *Parris 11440* (K), 1700 m, *11452* (K).

26.7.3. Metathelypteris gracilescens (Blume) Ching, Acta Phytotax. Sinica 8: 305 (1963).

Dryopteris gracilescens (Blume) Kuntze, Rev. Gen. Pl. 2: 812 (1891). C. Chr. & Holttum, Gardens' Bull. 7: 240 (1934).

Terrestrial. Lower montane forest. Elevation: 1400–2400 m.

Material examined: GIGISSEN CREEK/MARAI PARAI: 1400 m, *Clemens 32383A* (SING); KIAU VIEW TRAIL: 1600 m, *Parris 10792* (K), 1600 m, *10798* (K), 1600 m, *Parris & Croxall 8532* (K); POWER STATION: 1800 m, *Holttum 53* (K, SING); SUMMIT TRAIL: 2400 m, *Parris & Croxall 8665* (K), 2400 m, *8865* (K); TENOMPOK: 1400 m, *Holttum SFN 25384* (K, SING); ULAR HILL TRAIL: 1800 m, *Parris 11461* (K), 1800 m, *Parris & Croxall 9079* (K).

26.8. PARATHELYPTERIS

26.8.1. Parathelypteris beddomei (Baker) Ching, Acta Phytotax. Sinica 8: 302 (1963).

a. var. **beddomei**

Terrestrial. Hill dipterocarp forest to lower montane forest. Elevation: 600–2000 m.

Material examined: KEMBURONGOH/LUMU-LUMU: 2000 m, *Molesworth-Allen 3233* (US); LANGANAN FALLS: 1000 m, *Parris & Croxall 8931* (K); PORING HOT SPRINGS/LANGANAN WATER FALLS: 600–1000 m, *Kokawa & Hotta 4930* (K); ULAR HILL TRAIL: 1800 m, *Parris 11458* (K).

26.9. PLESIONEURON

26.9.1. Plesioneuron fuchsii Holttum, Blumea 22: 236 (1975). Type: Goking's Valley: 2700 m, *Fuchs 21477* (holotype L n.v.; isotype K!).

Terrestrial. Lower montane forest. Elevation: 2300–2700 m. Endemic to Mount Kinabalu.

Additional material examined: KILEMBUN BASIN: 2300 m, *Clemens 33719* (K, SING).

26.10. PNEUMATOPTERIS

26.10.1. Pneumatopteris callosa (Blume) Nakai, Bot. Mag. Tokyo 47: 179 (1933).

Dryopteris callosa (Blume) C. Chr., Index Fil., 256 (1905). C. Chr. & Holttum, Gardens' Bull. 7: 245 (1934).

Terrestrial. Lower montane forest. Elevation: 1200–1700 m.

Material examined: BAMBANGAN RIVER: 1500–1600 m, *Hotta 20318* (L); LUBANG: *Clemens 10347* (K); MAMUT RIVER: 1200 m, *RSNB 1678* (K, SING); MARAI PARAI: 1500 m, *Clemens 32594* (K); MESILAU RIVER: 1500 m, *RSNB 1373* (K, SING); MOUNT KINABALU: *Clemens 32593a* (L); PARK HEADQUARTERS TO POWER STATION: 1700 m, *Parris 11439* (K); PARK HEADQUARTERS: 1400 m, *Edwards 2208* (K), 1500 m, *Parris 10839* (K), 1500 m, *Parris & Croxall 8560* (K); TAHUBANG FALLS: 1500 m, *Clemens 40309* (K).

26.10.2. Pneumatopteris microauriculata Holttum, Blumea 21: 311 (1973). Type: Mount Kinabalu: *Clemens 27137* (holotype BM n.v.; isotype K!).

Terrestrial. Hill dipterocarp forest. Elevation: 700–900 m. Endemic to Mount Kinabalu.

Additional material examined: DALLAS: 900 m, *Clemens 30463* (K, SING); LANGANAN FALLS: 700 m, *Parris & Croxall 8949* (K).

26.10.3. Pneumatopteris micropaleata Holttum, Blumea 21: 319 (1973). Type: Mount Kinabalu: 1800 m, *Holttum 58* (holotype K!; isotype SING!).

Terrestrial. Lower montane forest. Elevation: 1800 m. Endemic to Mount Kinabalu.

26.10.4. Pneumatopteris truncata (Poir.) Holttum, Blumea 21: 314 (1973).

Dryopteris truncata (Poir.) C. Chr., Index Fil., 299 (1905). C. Chr. & Holttum, Gardens' Bull. 7: 246 (1934). *Pneumatopteris christelloides* Holttum, Blumea 21: 311 (1973). Type: Dallas: 900 m, *Clemens 27541* (holotype K!; isotype BM n.v.).

Terrestrial. Lowland dipterocarp forest to lower montane forest. Elevation: 500–1700 m.

Additional material examined: DALLAS: 900 m, *Clemens 27029* (K), 1100 m, *Holttum SFN 25260* (K, SING); EASTERN SHOULDER, CAMP 1: 1200 m, *RSNB 1214* (K, SING); KILEMBUN RIVER: 1400 m, *Clemens 32465* (SING); LIWAGU RIVER TRAIL: 1700 m, *Parris & Croxall 9101* (K); MARAI PARAI: 1400 m, *Clemens 32379* (US); MESILAU CAMP: 1500 m, *Holttum 24* (SING); MESILAU TRAIL: 1400–1600 m, *Kokawa & Hotta 4492* (US); PORING HOT SPRINGS: 500 m, *Parris & Croxall 8568* (K); PORING HOT SPRINGS/LANGANAN WATER FALLS: 600–900 m, *Kokawa & Hotta 4734* (K); TENOMPOK: 1500 m, *Clemens 28646* (K), 1500 m, *29450* (K).

26.11. PRONEPHRIUM

26.11.1. Pronephrium borneense (Hook.) Holttum, Fl. Males. 2, 1: 528 (1981).

Dryopteris labuanensis C. Chr., Index Fil., 273 (1905). C. Chr. & Holttum, Gardens' Bull. 7: 249 (1934).

Elevation: 900 m.

Material examined: DALLAS: 900 m, *Clemens 30458* (SING, US).

26.11.2. Pronephrium cuspidatum (Blume) Holttum, Blumea 20: 123 (1972).

Terrestrial. Lower montane forest. Elevation: 1200–1600 m.

Material examined: BUNDU TUHAN: 1400 m, *Carr s.n.* (SING); KIAU VIEW TRAIL: 1600 m, *Parris & Croxall 9126* (K); PARK HEADQUARTERS: 1500 m, *Parris 11622* (K); PENIBUKAN: 1200 m, *Clemens 40594* (K); TENOMPOK: 1400 m, *Holttum SFN 25410* (K, SING).

26.11.3. Pronephrium firmulum (Baker) Holttum, Blumea 20: 116 (1972).

Dryopteris firmula (Baker) C. Chr., Index Fil., 266 (1905). C. Chr. & Holttum, Gardens' Bull. 7: 249 (1934).

Terrestrial. Hill forest to lower montane forest. Elevation: 900–1400 m.

Material examined: DALLAS: 900 m, *Clemens 26923* (K), 900 m, *27430* (K), 1100 m, *Holttum SFN 25144* (K, SING); LIWAGU RIVER: 1400 m, *Parris & Croxall 9048* (K); PARK HEADQUARTERS: 1400 m, *Parris & Croxall 8603* (K); TAHUBANG RIVER: 900 m, *Holttum SFN 25596* (K, SING).

26.11.4. Pronephrium hosei (Baker) Holttum, Blumea 20: 120 (1972).

Dryopteris hosei (Baker) C. Chr., Index Fil., 271 (1905). C. Chr. & Holttum, Gardens' Bull. 7: 248 (1934).

Terrestrial. Lowland to lower montane forest? Elevation: 300–1200 m.

Material examined: KEBAYAU/KAUNG: 300–600 m, *Clemens 27663* (SING, US), 300–600 m, *27763* (K); NUNGKEK LUBANG: 1200 m, *Clemens 32485a* (K, SING); PORING HOT SPRINGS/LANGANAN WATER FALLS: 700 m, *Parris & Croxall 8955* (K).

26.11.5. Pronephrium menisciicarpon (Blume) Holttum, Blumea 20: 111 (1972).

Dryopteris mirabilis Copel., Philipp. J. Sci. 6C: 137, pl. 19 (1911). C. Chr. & Holttum, Gardens' Bull. 7: 249 (1934). *Dryopteris labuanensis* sensu C. Chr. & Holttum non C. Chr., Gardens' Bull. 7: 249 (1934).

Terrestrial. Montane dipterocarp forest. Elevation: 900–1400 m.

Material examined: DALLAS: 1100 m, *Holttum SFN 25147* (K, SING); LOHAN/MAMUT COPPER MINE: 900 m, *Beaman 10640* (MICH); PINOSUK PLATEAU: 1400 m, *Beaman 10727* (MICH, MSC).

26.11.6. Pronephrium nitidum (Holttum) Holttum, Blumea 20: 109 (1972).

Dryopteris urophylla (Mett.) C. Chr. var. *nitida* Holttum in C. Chr. & Holttum, Gardens' Bull. 7: 249 (1934). Type: Minitinduk: 900 m, *Holttum SFN 25592* (holotype SING!; isotypes BO n.v., K!).

Terrestrial. Lowland dipterocarp forest to lower montane forest. Elevation: 500–1600 m. Not endemic to Mount Kinabalu.

Additional material examined: KINATEKI RIVER: 900 m, *Carr SFN 26830* (SING); KIPUNGIT FALLS TRAIL: 600 m, *Parris & Croxall 8899* (K); MEMPENING TRAIL: 1600 m, *Parris & Croxall 9120* (K); MESILAU CAMP: 900 m, *Holttum 23* (SING); PENDIRUEN RIVER: 600 m, *Shea & Aban SAN 76906* (K); PORING HOT SPRINGS: 600 m, *Beaman 7575* (MICH, MSC); TAKUTAN: 500 m, *Shea & Aban SAN 77159* (K).

26.11.7. Pronephrium peltatum (Alderw.) Holttum

a. var. **aberrans** Holttum, Fl. Males. 2, 1: 530 (1981). Type: Kilembun Basin: *Clemens "33702 & 32020"* (holotype K!).

Terrestrial. Lower montane forest. Elevation: 1400–1800 m. Not endemic to Mount Kinabalu.

Additional material examined: SEDIKEN RIVER: 1500–1800 m, *Clemens 32299* (US).

b. var. **persetiferum** Holttum, Fl. Males. 2, 1: 530 (1981). Type: Penibukan: *Holttum s.n.* (holotype SING!; isotype K!).

Dryopteris exsculpta sensu C. Chr. & Holttum non (Baker) C. Chr., Gardens' Bull. 7: 248 (1934) p.p.

Terrestrial. Hill dipterocarp to lower montane forest. Elevation: 900–1600 m. Not endemic to Mount Kinabalu.

Additional material examined: NUNGKEK LUBANG: 1200 m, *Clemens 32485* (K, SING); PARK HEADQUARTERS: 1600 m, *Parris & Croxall 8539b* (K); PENIBUKAN: 1100 m, *Parris 11518* (K); PORING HOT SPRINGS/LANGANAN WATER FALLS: 1000 m, *Parris & Croxall 8922* (K).

c. var. **tenompokensis** (C. Chr.) Holttum, Fl. Males. 2, 1: 530 (1981).

Dryopteris tenompokensis C. Chr. in C. Chr. & Holttum, Gardens' Bull. 7: 248 (1934). Type: Tenompok: 1400 m, *Holttum SFN 25388* (holotype BM n.v.; isotypes K!, SING!).

Terrestrial. Lower montane forest. Elevation: 1400–1600 m. Endemic to Mount Kinabalu.

Additional material examined: PARK HEADQUARTERS: 1600 m, *Parris & Croxall 8539a* (K).

26.12. PSEUDOPHEGOPTERIS

26.12.1. Pseudophegopteris aurita (Hook.) Ching, Acta Phytotax. Sinica 8: 314 (1963).

Terrestrial. Lower montane forest. Elevation: 1200–1800 m.

Material examined: LIWAGU RIVER TRAIL: 1800 m, *Parris & Croxall 9105* (K); PENIBUKAN: 1200 m, *Clemens 40306* (K); ULAR HILL TRAIL: 1800 m, *Parris 11459* (K).

26.12.2. Pseudophegopteris kinabaluensis Holttum, Blumea 17: 16 (1969). Type: Goking's Valley: 2700 m, *Fuchs 21475* (holotype SSR n.v.; isotypes K!, L. n.v.).

Terrestrial. Lower montane forest. Elevation: 1800–2800 m. Endemic to Mount Kinabalu.

Additional material examined: GOKING'S VALLEY: 2800 m, *Fuchs 21484* (K); POWER STATION: 1800 m, *Holttum 57* (K, SING).

26.12.3. Pseudophegopteris paludosa (Blume) Ching, Acta Phytotax. Sinica 8: 315 (1963).

Terrestrial. Lower montane forest. Elevation: 1700–1800 m.

Material examined: KADAMAIAN FALLS TRAIL: 1700 m, *Parris & Croxall 9085* (K); ULAR HILL TRAIL: 1800 m, *Parris & Croxall 9084* (K).

26.12.4. Pseudophegopteris rectangularis (Zoll.) Holttum, Blumea 17: 19 (1969).

Terrestrial. Probably lower montane forest. Elevation: 900–1200 m.

Material examined: PENATARAN BASIN: 900–1200 m, *Clemens 34133* (K).

26.13. SPHAEROSTEPHANOS

26.13.1. Sphaerostephanos baramensis (C. Chr.) Holttum, Fl. Males. 2, 1: 473 (1981).

Dryopteris baramensis C. Chr. in C. Chr. & Holttum, Gardens' Bull. 7: 246 (1934).

Terrestrial. Lower montane forest. Elevation: 1400–1700 m.

Material examined: KIAU VIEW TRAIL: 1600 m, *Parris & Croxall 8534* (K); LIWAGU RIVER TRAIL: 1700 m, *Parris & Croxall 9102* (K); MESILAU CAMP: 1500 m, *Holttum 1* (SING); PARK HEADQUARTERS: 1400 m, *Edwards 2173* (K), 1500 m, *Parris 10837* (K); TENOMPOK: 1500 m, *Clemens 26932* (K), 1500 m, *27601* (K), 1500 m, *28453* (SING), 1500 m, *30456* (K), 1400 m, *Holttum SFN 25385* (K, SING).

26.13.2. Sphaerostephanos caulescens Holttum, Fl. Males. 2, 1: 472 (1981).

Dryopteris porphyricola sensu C. Chr. & Holttum non Copel., Gardens' Bull. 7: 244 (1934).

Terrestrial. Lower montane forest. Elevation: 1200–1800 m.

Material examined: KILEMBUN BASIN: 1200 m, *Clemens 33999* (K); KILEMBUN RIVER: 1400 m, *Clemens 32464* (K, SING), 1200 m, *33999* (SING); LIWAGU RIVER TRAIL: 1400 m, *Parris & Croxall 9050* (K), 1400 m, *9051* (K); MESILAU CAMP: 1500 m, *Holttum 19* (K, SING); TENOMPOK: 1500 m, *Clemens 27190* (K), 1500 m, *27471* (US), 1500 m, *28456* (K), 1400 m, *Holttum SFN 25389* (K, SING); WEST MESILAU RIVER: 1800 m, *Clemens 29054* (US).

26.13.3. Sphaerostephanos gymnorachis Holttum, Fl. Males. 2, 1: 479 (1981). Type: Pinosuk Plateau: 1500 m, *Holttum 34* (holotype K!; isotype SING!).

Dryopteris heterocarpa sensu C. Chr. & Holttum p.p. non (Blume) Kuntze, Gardens' Bull. 7: 244 (1934).

Terrestrial. Lower montane forest. Elevation: 1400–2100 m. Endemic to Mount Kinabalu.

Additional material examined: LIWAGU RIVER TRAIL: 1400 m, *Parris & Croxall 9049* (K); LUBANG: *Topping 1788* (SING); MESILAU BASIN: 2100 m, *Clemens 29050* (K, SING); TENOMPOK: 1500 m, *Clemens 28233* (K), 1500 m, *28282* (SING), 1500 m, *28382* (K).

26.13.4. Sphaerostephanos heterocarpus (Blume) Holttum in Nayar & Kaur, Comp. to Bedd., 209 (1974).

Terrestrial. Lowland dipterocarp forest to hill forest, frequently on ultramafics. Elevation: 500–1400 m.

Material examined: LOHAN/MAMUT COPPER MINE: 1000 m, *Beaman 10649* (MICH, MSC); PENIBUKAN: 1200 m, *Parris 11520* (K); PINOSUK PLATEAU: 1400 m, *Beaman 10725* (K, MICH, MSC); PORING HOT SPRINGS: 500 m, *Parris & Croxall 9001* (K); PORING HOT SPRINGS/LANGANAN WATER FALLS: 1000 m, *Parris & Croxall 8934* (K), 600 m, *8968* (K); TAKUTAN: 800 m, *Shea & Aban SAN 77241* (K, SING).

26.13.5. Sphaerostephanos inconspicuus (Copel.) Holttum, Fl. Males. 2, 1: 493 (1981).

Dryopteris inconspicua Copel., Philipp. J. Sci. 12C: 55 (1917). C. Chr. & Holttum, Gardens' Bull. 7: 242 (1934). Type: Kiau: *Topping 1543* (holotype PNH?†; isotype NY n.v.).

Terrestrial. Hill forest to lower montane forest. Elevation: 900–1800 m. Not endemic to Mount Kinabalu.

Material examined: DALLAS: 900 m, *Clemens 27603* (K); DALLAS/TENOMPOK: *Clemens 26893* (US); GURULAU SPUR: 1800 m, *Clemens 50578* (K), *Topping 1837* (US); KIAU VIEW TRAIL: 1600 m, *Parris 10790* (K), 1600 m, *10797* (K); LIWAGU RIVER TRAIL: 1500 m, *Parris 11486* (K), 1600 m, *Parris & Croxall 9093* (K); MARAI PARAI: 1500 m, *Clemens 33090* (US); MESILAU CAMP: 1500 m, *Holttum 22* (SING); PARK HEADQUARTERS: 1600 m, *Parris & Croxall 8538* (K); PENIBUKAN: 1200 m, *Clemens 40564* (K); TENOMPOK: 1500 m, *Clemens 26983* (SING), 1500 m, *28099* (K).

26.13.6. Sphaerostephanos latebrosus (Kunze ex Mett.) Holttum in Nayar & Kaur, Comp. to Bedd., 209 (1974).

Terrestrial. Hill dipterocarp forest. Elevation: 800–900 m.

Material examined: EASTERN SHOULDER: 800 m, *RSNB 576* (K, SING); EASTERN SHOULDER, CAMP 1: 900 m, *RSNB 1172* (K, SING).

26.13.7. Sphaerostephanos lithophyllus (Copel.) Holttum, Fl. Males. 2, 1: 496 (1981).

Dryopteris lithophylla Copel., Philipp. J. Sci. 12C: 57 (1917). C. Chr. & Holttum, Gardens' Bull. 7: 245 (1934). Type: Marai Parai Spur: *Topping 1850½* (holotype MICH!; isotype US!).

Terrestrial. Lower montane forest on ultramafics. Elevation: 1500–2200 m. Endemic to Mount Kinabalu.

Additional material examined: BAMBANGAN RIVER: 1500 m, *RSNB 1310* (K, SING); MARAI PARAI: 1500 m, *Clemens 33158* (K), 1500 m, *Holttum SFN 25606* (K, SING); MESILAU CAVE: 1900–2200 m, *Beaman 9541* (K, MICH, MSC); MESILAU RIVER: 2100 m, *Collenette 21612* (K).

26.13.8. Sphaerostephanos lobangensis (C. Chr.) Holttum, Fl. Males. 2, 1: 479 (1981).

Dryopteris lobangensis C. Chr. in C. Chr. & Holttum, Gardens' Bull. 7: 245 (1934). Type: Lubang/Paka-paka Cave: *Clemens 10728* (holotype MICH!; isotype K!).

Terrestrial. Lower montane forest on ultramafics. Elevation: 2600–2700 m. Endemic to Mount Kinabalu.

Additional material examined: SUMMIT TRAIL: 2700 m, *Parris 11559* (K), 2600 m, *Parris & Croxall 8696* (K).

26.13.9. Sphaerostephanos neotoppingii Holttum, Fl. Males. 2, 1: 500 (1981).

Dryopteris toppingii Copel., Philipp. J. Sci. 7: 246 (1934). C. Chr. & Holttum, Gardens' Bull. 7: 246 (1934). Type: Lubang: *Topping 1766* (isotype MICH!).

Terrestrial. Hill dipterocarp forest and lower montane forest. Elevation: 900–1700 m. Not endemic to Mount Kinabalu.

Additional material examined: DALLAS: 900 m, *Clemens 27543* (K); DALLAS/TENOMPOK: 1200 m, *Clemens 27729* (K); EASTERN SHOULDER, CAMP 1: 1200 m, *RSNB 1213* (K, SING); PINOSUK PLATEAU: 1500 m, *Holttum 37* (K, SING), 1500 m, *Parris & Croxall 9145* (K); TENOMPOK: 1700 m, *Clemens 28082* (K), 1400 m, *Holttum SFN 25395* (K, SING).

26.13.10. Sphaerostephanos penniger (Hook.) Holttum in Nayar & Kaur, Comp. to Bedd., 209 (1974).

a. var. **penniger**

Dryopteris megaphylla (Mett.) C. Chr., Index Fil., 277 (1905). C. Chr. & Holttum, Gardens' Bull. 7: 246 (1934) p.p.

Terrestrial. Hill dipterocarp forest to lower montane forest? Elevation: 800–1200 m.

Material examined: DALLAS: 1100 m, *Clemens 26766* (US), 800 m, *27288* (US), 900 m, *27317* (US); LUBANG: *Topping 1788* (US); MARAI PARAI: 1200 m, *Clemens 32398* (K, SING).

26.13.11. Sphaerostephanos perglanduliferus (Alderw.) Holttum, Kalikasan 4: 59 (1975).

Dryopteris megaphylla sensu C. Chr. & Holttum p.p. non (Mett.) C. Chr., Gardens' Bull. 7: 246 (1934).

Terrestrial. Lowland dipterocarp forest. Elevation: 500–1000 m.

Material examined: KIAU: *Topping 1531* (US); PINAWANTAI: 600 m, *Shea & Aban SAN 76911* (K); PORING HOT SPRINGS: 500 m, *Parris & Croxall 9004* (K); PORING HOT SPRINGS/LANGANAN WATER FALLS: 600–1000 m, *Kokawa & Hotta 4922* (K).

26.13.12. Sphaerostephanos polycarpus (Blume) Copel., Univ. Calif. Publ. Bot. 16: 60 (1929).

Mesochlaena toppingii Copel., Philipp. J. Sci. 12C: 57 (1917). C. Chr. & Holttum, Gardens' Bull. 7: 254 (1934). Type: Kebayau/Kaung: *Topping 1902* (holotype MICH!).

Terrestrial. Hill forest. Elevation: 600–900 m.

Additional material examined: DALLAS: 900 m, *Clemens 26908* (K), 900 m, *27458* (K); KAUNG/DALLAS: 600 m, *Holttum SFN 25130* (K, SING).

26.13.13. Sphaerostephanos porphyricola (Copel.) Holttum, Kalikasan 4: 59 (1975).

Terrestrial. Lowland dipterocarp forest. Elevation: 500–600 m.

Material examined: KIPUNGIT FALLS TRAIL: 600 m, *Parris & Croxall 8902* (K); PORING HOT SPRINGS: 500 m, *Parris & Croxall 9014* (K).

26.13.14. Sphaerostephanos squamatellus Holttum, Fl. Males. 2, 1: 497 (1981). Type: Power Station: 1900 m, *Holttum 56* (holotype K!; isotype SING!).

Terrestrial. Lower montane forest. Elevation: 1500–2400 m. Endemic to Mount Kinabalu.

Additional material examined: PENATARAN RIVER: 1500 m, *Clemens 32548A* (SING); SUMMIT TRAIL: 2400 m, *Parris 11610* (K), 2300 m, *Parris & Croxall 8864* (K); ULAR HILL TRAIL: 1800 m, *Parris & Croxall 9080* (K).

26.13.15. Sphaerostephanos trichochlamys Holttum, Fl. Males. 2, 1: 465 (1981). Type: Mount Kinabalu: *Holttum 4* (holotype K!).

Dryopteris heterocarpa sensu C. Chr. & Holttum p.p. non (Blume) Kuntze, Gardens' Bull. 7: 244 (1934).

Terrestrial. Lower montane forest. Elevation: 1200–2300 m. Not endemic to Mount Kinabalu.

Additional material examined: GIGISSEN CREEK/MARAI PARAI: 1400 m, *Clemens 32393* (K, SING); LIWAGU RIVER TRAIL: 1500 m, *Parris 10828* (K); MARAI PARAI: 1500 m, *Clemens 32593* (K, SING); PARK HEADQUARTERS: 1500 m, *Parris 10835* (K), 1500 m, *Parris & Croxall 8536* (K); PENIBUKAN: 1200 m, *Clemens 30553* (K), 1200 m, *30736* (K), 1200 m, *Holttum SFN 25600* (K, SING); SUMMIT TRAIL: 2300 m, *Parris & Croxall 8663* (K); ULAR HILL TRAIL: 1700 m, *Parris 11454* (K).

26.13.16. Sphaerostephanos unitus (L.) Holttum

a. var. **mucronatus** (Christ) Holttum, Fl. Males. 2, 1: 478 (1981).

Dryopteris unita (L.) Kuntze, Rev. Gen. Pl. 2: 811 (1891). C. Chr. & Holttum, Gardens' Bull. 7: 245 (1934).

Terrestrial. Lowland forest to lower montane forest. Elevation: 1100–1500 m.

Material examined: DALLAS: 1100 m, *Holttum SFN 25263* (SING); MEKEDEU RIVER: *Shea & Aban SAN 77269* (K); TENOMPOK: 1500 m, *Clemens 26928* (K), 1500 m, *29210* (K).

26.14. STEGNOGRAMMA

26.14.1. Stegnogramma aspidioides Blume, Enum. Pl. Javae, 173 (1828).

Dryopteris stegnogramma (Blume) C. Chr., Index Fil., 294 (1905). C. Chr. & Holttum, Gardens' Bull. 7: 250 (1934).

Terrestrial. Lower montane forest. Elevation: 2100 m.

Material examined: KEMBURONGOH: 2100 m, *Holttum SFN 25538* (SING); MOUNT KINABALU: *Holttum s.n.* (K).

27. VITTARIACEAE

27.1. ANTROPHYUM

Copeland, E. B. 1960. Fern Flora of the Philippines. Vol. 3: 543–547.

27.1.1. Antrophyum callifolium Blume, Enum. Pl. Javae, 111 (1828). C. Chr. & Holttum, Gardens' Bull. 7: 315 (1934).

Antrophyum callifolium Blume var. *magnum* C. Chr. in C. Chr. & Holttum, Gardens' Bull. 7: 315 (1934).

Epiphyte. Hill forest. Elevation: 800–900 m.

Material examined: DALLAS: 900 m, *Clemens 26384* (K), 900 m, *27137* (K), 900 m, *27454* (K), 900 m, *27492* (K), 800 m, *Holttum SFN 25362* (K, SING); EASTERN SHOULDER: 900 m, *RSNB 123* (K, SING).

27.1.2. Antrophyum coriaceum (D. Don) Wall., List no. 43 (1828). C. Chr. & Holttum, Gardens' Bull. 7: 315 (1934).

Epiphyte. Hill forest. Elevation: 800 m.

Material examined: DALLAS: 800 m, *Clemens 27441* (BM).

27.1.3. Antrophyum latifolium Blume, Fl. Javae 2: 75 (1828).

Epiphyte. Lower montane forest. Elevation: 1500–1800 m.

Material examined: PENIBUKAN: 1500–1800 m, *Clemens 40294* (K, L), 1500–1800 m, *40963* (L).

27.1.4. Antrophyum lessonii Bory, Dup. Voy. Bot. 1: 255, t. 28, f. 2 (1828). C. Chr. & Holttum, Gardens' Bull. 7: 315 (1934).

Lithophyte and epiphyte. Hill forest. Elevation: 900 m.

Material examined: DALLAS: 900 m, *Clemens 27402* (SING).

27.1.5. Antrophyum parvulum Blume, Enum. Pl. Javae 110 (1828). C. Chr. & Holttum, Gardens' Bull. 7: 315 (1934).

Epiphyte. Elevation: 1500 m.

Material examined: TENOMPOK: 1500 m, *Clemens s.n.* (BM).

27.1.6. Antrophyum reticulatum (G. Forster) Kaulf., Enum., 197 (1824).

Epiphyte. Hill dipterocarp forest and lower montane forest. Elevation: 600–1600 m.

Material examined: MELANGKAP KAPA: 600–700 m, *Beaman 8578* (K, MICH, MSC), 700–1000 m, *8780* (MICH); WEST MESILAU RIVER: 1600 m, *Beaman 9030* (K, MICH, MSC).

27.1.7. Antrophyum semicostatum Blume, Enum. Pl. Javae, 110 (1828). C. Chr. & Holttum, Gardens' Bull. 7: 315 (1934).

Epiphyte. Hill forest and lower montane forest. Elevation: 600–2100 m.

Material examined: DALLAS: 900 m, *Clemens 27455* (K), 1100 m, *Holttum SFN 25150* (SING); KEMBURONGOH: 2100 m, *Holttum SFN 25536* (SING); LIWAGU RIVER: 1400 m, *Parris & Croxall 8607* (K); MAMUT RIVER: 1200 m, *RSNB 1709* (K, SING); PENIBUKAN: 1200 m, *Clemens 30611* (K);

PINAWANTAI: 600 m, *Shea & Aban SAN 76909* (K); TENOMPOK: 1500 m, *Clemens 29417* (K); WEST MESILAU RIVER: 1600 m, *Beaman 9357* (K, MICH, MSC).

27.1.8. Antrophyum sessilifolium (Cav.) Spreng., Syst. 4: 67 (1827).

Epiphyte and lithophyte. Lowland dipterocarp forest and lower montane forest. Elevation: 500–1500 m.

Material examined: KIPUNGIT FALLS: 600 m, *Parris & Croxall 8914* (K), 600 m, *8965* (K); PARK HEADQUARTERS: 1500 m, *Parris 11427* (K); PENATARAN RIVER: 500 m, *Beaman 9322* (MICH, MSC).

27.2. VAGINULARIA

Tagawa, M. & Iwatsuki, K. 1985. Flora of Thailand. Vol. 3 (2): 228–230.

27.2.1. Vaginularia paradoxa (Fée) Mett., Ann. Lugd. Bat. 4: 174 (1868–1869).

Monogramma trichoidea sensu C. Chr. & Holttum p.p. non (Fée) Hook. & Baker, Gardens' Bull. 7: 316 (1934).

Epiphyte. Lower montane forest. Elevation: 1400–1500 m.

Material examined: PARK HEADQUARTERS: 1500 m, *Parris & Croxall 8589* (K); TENOMPOK: 1400 m, *Holttum SFN 25715* (SING).

27.2.2. Vaginularia trichoidea Fée, Mem.3, 54 (1851–52).

Monogramma trichoidea (Fée) Hook. & Baker, Syn. Fil., 375 (1865–1868). C. Chr. & Holttum, Gardens' Bull. 7: 316 (1934) p.p.

Epiphyte. Lower montane forest. Elevation: 1500 m.

Material examined: TENOMPOK: 1500 m, *Clemens 29574* (K).

27.3. VITTARIA

Copeland, E. B. 1960. Fern Flora of the Philippines. Vol. 3: 547–553. Holttum, R. E. 1968. Revised Flora of Malaya. Vol. 2, Ferns. Ed. 2: 607–614.

27.3.1. Vittaria angustifolia Blume, Enum. Pl. Javae, 199 (1828). C. Chr. & Holttum, Gardens' Bull. 7: 316 (1934).

Epiphyte. Lower montane forest. Elevation: 1400–2300 m.

Material examined: KEMBURONGOH: 2300 m, *Clemens 27953* (K); KEMBURONGOH/LUMU-LUMU: 1500 m, *Sinclair et al. 9026* (K, SING); KIAU VIEW TRAIL: 1600 m, *Parris & Croxall 9134* (K); KINATEKI RIVER: 1500 m, *Clemens 31059* (K); LIWAGU RIVER TRAIL: 1700 m, *Parris 11473* (K), 1400 m, *Parris & Croxall 8590* (K); TENOMPOK/LUMU-LUMU: 1500 m, *Holttum SFN 25437* (K).

27.3.2. Vittaria elongata Sw., Syn., 109, 302 (1806). C. Chr. & Holttum, Gardens' Bull. 7: 316 (1934) p.p.

Epiphyte. Lowland dipterocarp forest and lower montane forest. Elevation: 600–1400 m.

Material examined: PORING HOT SPRINGS/LANGANAN WATER FALLS: 600 m, *Parris & Croxall 8954* (K); TENOMPOK: 1400 m, *Holttum SFN 25419* (K, SING).

27.3.3. Vittaria ensiformis Sw., Ges. Nat. Fr. Berl. Neu. Schr. 2: 134, t. 7, f. 1 (1799). C. Chr. & Holttum, Gardens' Bull. 7: 316 (1934).

Epiphyte. Hill forest and lower montane forest. Elevation: 900–1500 m.

Material examined: DALLAS: 900 m, *Clemens 27443* (K); TENOMPOK: 1500 m, *Clemens 28185* (K).

27.3.4. Vittaria incurvata Cav., Descr., 270 (1802).

Epiphyte. Hill dipterocarp forest. Elevation: 1300 m.

Material examined: KIAU: *Topping 1552* (SING, US); LUGAS HILL: 1300 m, *Beaman 10569* (MICH, MSC).

27.3.5. Vittaria lineata (L.) Sm.?, Mem. Ac. Turin. 5: 42, t. 9, f. 5 (1793). C. Chr. & Holttum, Gardens' Bull. 7: 316 (1934).

Epiphyte? Based on *Clemens 10007,* Kiau (BM? n.v.); *Haslam* (BM? n.v.).

27.3.6. Vittaria parvula Bory, Bel. Voy. Bot. 2: 35 (1833).

Epiphyte. Lower montane forest. Elevation: 2300 m.

Material examined: SUMMIT TRAIL: 2300 m, *Parris & Croxall 8667* (K).

27.3.7. Vittaria zosterifolia Willd., Sp. Pl. 5: 406 (1810).

Vittaria elongata sensu C. Chr. & Holttum p.p. non Sw., Gardens' Bull. 7: 316 (1934).

Epiphyte. Hill forest to lower montane forest. Elevation: 900–1500 m.

Material examined: DALLAS: 900 m, *Clemens 27512* (K); GURULAU SPUR: *Topping 1833* (SING, US); KEBAYAU/KAUNG: *Topping 1908* (US); KIAU/LUBANG: *Topping 1612* (US); LUBANG: *Clemens 10325* (K); TENOMPOK: 1500 m, *Clemens 29581* (K, US).

28. WOODSIACEAE

28.1. ACYSTOPTERIS

Holttum, R. E. 1968. Revised Flora of Malaya. Vol. 2, Ferns. Ed. 2: 540–541.

28.1.1. Acystopteris tenuisecta (Blume) Tagawa, Acta Phytotax. Geobot. 7: 73 (1938).

Dryopteris dennstaedtioides Copel., Philipp. J. Sci. 56: 473, pl. 3 (1935). Type: Tinekuk Falls: *Clemens 50032* (holotype PNH?†).

Terrestrial. Lower montane slope and valley oak forest. Elevation: 1500–2700 m.

Material examined: KILEMBUN BASIN: 2700 m, *Clemens 33748* (K); LIWAGU RIVER TRAIL: 1500 m, *Parris 11490* (K); SUMMIT TRAIL: 1800 m, *Parris & Croxall 8872* (K); ULAR HILL TRAIL: 1700 m, *Parris 11448* (K).

28.2. ATHYRIUM

Copeland, E. B. 1917. Keys to the ferns of Borneo. Sarawak Mus. J. 2: 374–380 [most species now belong in *Diplazium*]. Tagawa, M. & Iwatsuki, K. 1988. Flora of Thailand. Vol. 3 (3): 445–449.

28.2.1. Athyrium amoenum C. Chr. in C. Chr. & Holttum, Gardens' Bull. 7: 267, pl. 56 (1934). Type: Kemburongoh: 2100 m, *Holttum SFN 25530* (holotype BM n.v.; isotypes SING!, US!).

Terrestrial. Lower and upper montane forest. Elevation: 2100–3400 m. Endemic to Mount Kinabalu.

Additional material examined: GURULAU SPUR: 3400 m, *Clemens 50907* (K); KILEMBUN BASIN: 2600–2700 m, *Clemens 33745* (US); LAYANG-LAYANG/PAKA-PAKA CAVE: 3000 m, *Edwards 2191* (K); LUMU-LUMU: 2100 m, *Clemens 29967* (SING); MESILAU BASIN: 2100 m, *Clemens 29045* (K); PAKA-PAKA CAVE: 2700 m, *Meijer SAN 22066* (K), 3000 m, *Sinclair et al. 9170* (K, SING), *Topping 1716* (SING); PANAR LABAN: 3100 m, *Parris 11570* (K); SUMMIT TRAIL: 2600 m, *Parris & Croxall 8812* (K).

28.2.2. Athyrium anisopterum Christ, Bull. Boiss. 6: 962 (1898).

Athyrium macrocarpum sensu C. Chr. & Holttum non (Blume) Bedd., Gardens' Bull. 7: 266 (1934).

Terrestrial. Lower montane oak forest. Elevation: 1400–3000 m.

Material examined: EASTERN SHOULDER: 3000 m, *RSNB 883* (K); KEMBURONGOH: 2100 m, *Holttum SFN 25526* (K, SING); KILEMBUN BASIN: 2700–3000 m, *Clemens 31798* (K); KILEMBUN RIVER HEAD: 1400 m, *Clemens 32518* (SING); KINATEKI RIVER: 1500–2700 m, *Clemens 31959* (K); LUMU-LUMU: 2100 m, *Clemens 27947* (US); MARAI PARAI: 1800 m. *Holttum s.n.* (SING); MESILAU BASIN: 2100 m, *Clemens 29033* (K); SUMMIT TRAIL: 2400 m, *Parris 11542* (K), 2400 m, *Parris & Croxall 8678* (K), 2600 m, *8697* (K).

28.2.3. Athyrium clemensiae Copel., Philipp. J. Sci. 12C: 58 (1917). C. Chr. & Holttum, Gardens' Bull. 7: 266 (1934). Type: Summit Area: 4100 m, *Clemens 10621* (holotype PNH?†; isotype MICH!).

Terrestrial. Upper montane forest, *Leptospermum* scrub. Elevation: 3200–4100 m. Endemic to Mount Kinabalu.

Additional material examined: MOUNT KINABALU: 3700 m, *Holttum SFN 25486* (K, SING); PANAR LABAN: 3300 m, *Parris 11593* (K); SUMMIT AREA: *Clemens 27049* (US), 3700–4000 m, *27782* (US); SUMMIT TRAIL: 3200 m, *Parris & Croxall 8739* (K), 3500 m, *8779* (K), *Topping 1698* (US); UPPER KINABALU: 4100 m, *Clemens 27007* (K), 4100 m, *27182* (K), 4100 m, *27944* (K).

28.2.4. Athyrium pulcherrimum Copel., Philipp. J. Sci. 8C: 141, pl. 3 (1913). C. Chr. & Holttum, Gardens' Bull. 7: 267 (1934).

Terrestrial. Lower montane forest. Elevation: 2400–3000 m.

Material examined: PAKA-PAKA CAVE: 3000 m, *Clemens 30446* (K), 3000 m, *Holttum SFN 25507* (K, SING), *Topping 1716* (US); SUMMIT TRAIL: 2500 m, *Parris 11608* (K), 2400 m, *Parris & Croxall 8862* (K).

28.2.5. Athyrium sp. 1

Terrestrial. Lower montane oak forest. Elevation: 1700–2400 m. Endemic to Mount Kinabalu.

Material examined: SUMMIT TRAIL: 2400 m, *Parris & Croxall 8861* (K); ULAR HILL TRAIL: 1700 m, *Parris 11451* (K).

28.3. CORNOPTERIS

Tagawa, M. & Iwatsuki, K. 1988. Flora of Thailand. Vol. 3 (3): 441–442.

28.3.1. Cornopteris opaca (D. Don) Tagawa, Acta Phytotax. Geobot. 8: 92 (1939).

Terrestrial. Lower montane oak forest. Elevation: 1600–1800 m.

Material examined: LIWAGU RIVER TRAIL: 1600 m, *Parris 11479* (K); SUMMIT TRAIL: 1800 m, *Parris & Croxall 8874* (K), 1800 m, *8875* (K).

28.4. DEPARIA

Kato, M. 1984. A taxonomic study of the Athyrioid fern genus *Deparia*, with main reference to the Pacific species. J. Fac. Sci. Univ. Tokyo 13: 375–430.

28.4.1. Deparia biserialis (Baker) M. Kato, J. Fac. Sci. Univ. Tokyo 3, 13: 413 (1984).

Asplenium biseriale Baker in Stapf, Trans. Linn. Soc. Bot. 4: 252 (1894). Type: Tahubang River: 900 m, *Haviland 1475* (holotype K!). *Diplazium grammitoides* sensu C. Chr. & Holttum non C. Presl, Gardens' Bull. 7: 269 (1934).

Lithophytic. Rheophyte in lower montane forest. Elevation: 900–2100 m. Endemic to Mount Kinabalu.

Additional material examined: KADAMAIAN RIVER NEAR MINITINDUK: 900 m, *Holttum SFN 25577* (K); KIAU/LUBANG: *Topping 1587* (US); LIWAGU RIVER: 1400 m, *Parris & Croxall 8595* (K); LIWAGU RIVER TRAIL: 1400 m, *Edwards 2179* (K), 1500 m, *Parris 11494* (K); MAMUT RIVER: 1200 m, *RSNB 1232* (K, SING); MARAI PARAI SPUR: *Clemens 11011* (K); MESILAU BASIN: 2100 m, *Clemens 29052* (K); PENATARAN BASIN: 900 m, *Clemens 34044* (SING), 1500 m, *34138* (K); PINOSUK PLATEAU: 1700 m, *Fuchs & Collenette 21668* (K); WASAI RIVER (PENATARAN BASIN): 1100 m, *Clemens 32571* (K).

28.4.2. Deparia boryana (Willd.) M. Kato, Bot. Mag. Tokyo 90: 36 (1977).

Dryopteris boryana (Willd.) C. Chr., Index Fil., 255 (1905). C. Chr. & Holttum, Gardens' Bull. 7: 254 (1934).

Terrestrial. Hill forest. Elevation: 900 m.

Material examined: KINATEKI RIVER (NEAR MINITINDUK): 900 m, *Holttum SFN 25583* (K).

28.4.3. Deparia petersenii (Kunze) M. Kato, Bot. Mag. Tokyo 90: 37 (1977).

a. var. petersenii

Lower montane forest. Elevation: 1500 m.

Material examined: LIWAGU RIVER TRAIL: 1500 m, *Parris & Croxall 9094* (K).

28.5. DIPLAZIOPSIS

Copeland, E. B. 1960. Fern Flora of the Philippines. Vol. 3: 420.

28.5.1. Diplaziopsis javanica (Blume) C. Chr., Index Fil., 227 (1905). C. Chr. & Holttum, Gardens' Bull. 7: 275 (1934).

Terrestrial. Lower montane forest. Elevation: 900–1600 m.

Material examined: EASTERN SHOULDER: 1200 m, *RSNB 1533* (K, SING); LIWAGU RIVER: 1600 m, *Parris & Croxall 9099* (K); MINITINDUK: 900 m, *Holttum SFN 25581* (SING).

28.6. DIPLAZIUM

Copeland, E. B. 1917. Keys to the ferns of Borneo. Sarawak Mus. J. 2: 374–380 [as *Athyrium*]. Copeland, E. B. 1960. Fern Flora of the Philippines. Vol. 3: 379–418. Holttum, R. E. 1968. Revised Flora of Malaya. Vol. 2, Ferns. Ed. 2: 541–574. Parris, B. S., Jermy, A. C., Camus, J. M. & Paul, A. M. 1984. The Pteridophyta of Gunung Mulu National Park. *In* Studies on the Flora of Gunung Mulu National Park, Sarawak. Ed. A. C. Jermy: 216–218, Forest Dept., Kuching, Sarawak. Price, M. G. 1983. Several unusual Malesian *Diplazia*. Gardens' Bull. 36: 25–29.

28.6.1. Diplazium allantoideum M. G. Price, Contr. Univ. Mich. Herb. 16: 195 (1987).

Diplazium sylvaticum sensu C. Chr. & Holttum non (Bory) Sw., Gardens' Bull. 7: 268 (1934).

Terrestrial. Lowland dipterocarp to hill forest. Elevation: 600–1200 m.

Material examined: LUBANG: 1200 m, *Holttum SFN 25562* (K, SING); PORING HOT SPRINGS/LANGANAN WATER FALLS: 600 m, *Parris & Croxall 8956* (K).

28.6.2. Diplazium amplissimum (Baker) Diels, Nat. Pfl. 1(4): 227 (1899).

Terrestrial, fronds to 2 m long. Lower montane valley oak forest. Elevation: 1100–1600 m.

Material examined: KILEMBUN BASIN: 1100 m, *Clemens 33747* (SING); LIWAGU RIVER TRAIL: 1600 m, *Parris 11484* (K); PENATARAN BASIN: *Clemens 32563* (L, SING).

28.6.3. Diplazium atrosquamosum (Copel.) C. Chr. in C. Chr. & Holttum, Gardens' Bull. 7: 274 (1934).

Athyrium atrosquamosum Copel., Philipp. J. Sci. 12C: 59 (1917). Type: Marai Parai Spur: *Clemens 11051* (holotype PNH?†; isotype MICH!; photo of isotype at UC, K!).

Terrestrial. Lower montane forest. Elevation: 1500–2100 m. Endemic to Mount Kinabalu.

Additional material examined: LUMU-LUMU: 2100 m, *Clemens 29766* (K); MARAI PARAI: 1500 m, *Clemens 33163* (K); TENOMPOK/LUMU-LUMU: 1500 m, *Holttum SFN 25429* (K, SING).

28.6.4. Diplazium barbatum C. Chr. in C. Chr. & Holttum, Gardens' Bull. 7: 272, pl. 59 (1934). Type: Tenompok: 1400 m, *Holttum SFN 25386* (holotype BM n.v.; isotypes K!, SING!).

Terrestrial. Lower montane forest. Elevation: 1400–2100 m. Not endemic to Mount Kinabalu.

Additional material examined: MARAI PARAI: 1800–2100 m, *Clemens 32607* (K).

28.6.5. Diplazium beamanii M. G. Price, Contr. Univ. Mich. Herb. 16: 193, f. 1–3 (1987). Type: Pinosuk Plateau: 1400 m, *Beaman 10724* (holotype MICH!; isotypes K!, MSC!).

Low tree fern. Lower montane forest by stream. Elevation: 1400 m. Endemic to Mount Kinabalu.

28.6.6. Diplazium christii C. Chr., Index Fil., 229 (1905). C. Chr. & Holttum, Gardens' Bull. 7: 270 (1934).

Material examined: GURULAU SPUR: *Clemens 10860* (MICH).

28.6.7. Diplazium cordifolium Blume, Enum. Pl. Javae, 190 (1828). C. Chr. & Holttum, Gardens' Bull. 7: 274 (1934).

a. var. **cordifolium**

Terrestrial. Hill dipterocarp and lower montane forest. Elevation: 900–2400 m.

Material examined: DALLAS: 900 m, *Clemens 27385* (K), 900 m, *Holttum SFN 25139* (K, SING), 1100 m, *25140* (SING); DALLAS/LUMU-LUMU: 1100–2000 m, *Clemens 26898* (K); GOLF COURSE SITE: 1800 m, *Beaman 7478* (MICH, MSC); KIAU: *Topping 1550* (SING, US); KIAU VIEW TRAIL: 1600 m, *Parris & Croxall 9131* (K), 1600 m, *9132* (K); KILEMBUN BASIN: 2400 m, *Clemens 33810* (US); LUGAS HILL: 1300 m, *Beaman 10548* (K, MICH, MSC); LUMU-LUMU: 2000 m, *Clemens 27062* (US); TENOMPOK: 1500 m, *Clemens 26950* (K), 1500 m, *27502* (K), 1500 m, *29422* (K).

b. var. **pariens** (Copel.) C. Chr., Gardens' Bull. 7: 274 (1934).

Terrestrial. Lower montane oak forest. Elevation: 1200–1800 m.

Material examined: DALLAS/TENOMPOK: 1200 m, *Clemens 27433* (US), 1200 m, *28338* (US); GURULAU SPUR: *Topping 1631* (US), *1835* (SING, US); KIAU VIEW TRAIL: 1600 m, *Parris 10794* (K), 1600 m, *Parris & Croxall 8547* (K), 1600 m, *9130* (K); LUMU-LUMU: 1800 m, *Clemens 27066* (US); PARK HEADQUARTERS: 1500 m, *Parris 10836* (K); SUMMIT TRAIL: 1800 m, *Parris & Croxall 8886* (K).

28.6.8. Diplazium crenatoserratum (Blume) T. Moore, Index, 121, 325 (1859). C. Chr. & Holttum, Gardens' Bull. 7: 269 (1934).

Terrestrial. Hill dipterocarp and lower montane forest. Elevation: 900–1200 m.

Material examined: DALLAS: 900 m, *Clemens 28155* (K); MARAI PARAI SPUR: *Clemens 10914* (K), *Topping 1843* (US); PENIBUKAN: 1200 m, *Clemens 40565* (K); PORING HOT SPRINGS/LANGANAN WATER FALLS: 900 m, *Parris & Croxall 8946* (K).

28.6.9. Diplazium dilatatum Blume, Enum. Pl. Javae, 194 (1828). C. Chr. & Holttum, Gardens' Bull. 7: 273 (1934) p.p.

Terrestrial. Lowland dipterocarp and hill forest. Elevation: 600–1200 m.

Material examined: KIPUNGIT FALLS: 600 m, *Parris & Croxall 8913* (K); LUBANG: *Topping 1784* (SING); MINITINDUK: 900–1200 m, *Clemens 29618* (SING).

28.6.10. Diplazium dolichosorum Copel., Philipp. J. Sci. 1, Suppl.: 151 (1906).

Tree fern. Hill dipterocarp forest. Elevation: 1000 m.

Material examined: LOHAN/MAMUT COPPER MINE: 1000 m, *Beaman 10641* (K, MICH, MSC).

28.6.11. Diplazium falcinellum C. Chr. in C. Chr. & Holttum, Gardens' Bull. 7: 270 (1934).

Terrestrial. Lower montane oak forest. Elevation: 1400–1600 m.

Material examined: MEMPENING TRAIL: 1600 m, *Parris 11507* (K), 1600 m, *Parris & Croxall 9122* (K); PARK HEADQUARTERS: 1400 m, *Edwards 2174* (K); TENOMPOK: 1500 m, *Clemens 28412* (K, SING), 1500 m, *29275* (SING, US).

28.6.12. Diplazium forbesii (Baker) C. Chr., Index Fil. 232 (1905).

Terrestrial. Lower montane oak forest. Elevation: 1600 m. This may not be distinct from *D. cordifolium* var. *pariens*.

Material examined: MENTEKI RIVER: 1600 m, *Beaman 10775* (MICH, MSC).

28.6.13. Diplazium fuliginosum (Hook.) M. G. Price, Brit. Fern Gaz. 10: 260 (1970).

Asplenium fuliginosum Hook., Sp. Fil. 3: 120 (1859). C. Chr. & Holttum, Gardens' Bull. 7: 280 (1934). Type: Mount Kinabalu: *Low s.n.* (lectotype of Parris, here designated, K!; isolectotype CGE!).

Terrestrial. Lower montane forest. Elevation: 1600–2900 m. Not endemic to Mount Kinabalu.

Additional material examined: EASTERN SHOULDER: 2900 m, *RSNB 794* (K); GURULAU SPUR: 2400 m, *Clemens 50860* (K); KEMBURONGOH: 2100 m, *Holttum SFN 25529* (K, SING); KINATEKI/KILEMBUN RIVERS: 2100–2400 m, *Clemens 33723* (K); LIWAGU RIVER TRAIL: 1600 m, *Parris & Croxall 9104* (K); MESILAU BASIN: 2100–2400 m, *Clemens 29749* (US); SUMMIT TRAIL: 2200–2500 m, *Jacobs 5780* (K).

28.6.14. Diplazium hewittii (Copel.) C. Chr., Index Fil. Suppl. 1: 26 (1913). C. Chr. & Holttum, Gardens' Bull. 7: 273 (1934).

Terrestrial. Lower montane forest. Elevation: 1500 m.

Material examined: MARAI PARAI SPUR: 1500 m, *Holttum SFN 25616* (K, SING).

28.6.15. Diplazium laevipes C. Chr. in C. Chr. & Holttum, Gardens' Bull. 7: 271, pl. 58 (1934). Type: Dallas: 1100 m, *Holttum SFN 25259* (holotype BM n.v.; isotypes K!, SING!).

Terrestrial. Hill forest. Elevation: 1100 m. Not endemic to Mount Kinabalu.

Additional material examined: KIAU: *Topping 1537* (SING, US).

28.6.16. Diplazium latifolium T. Moore, Index Fil., 141 (1859).

Diplazium dilatatum sensu C. Chr. & Holttum p.p. non Blume, Gardens' Bull. 7: 273 (1934).

Terrestrial. Lower montane forest. Elevation: 1200–1500 m.

Material examined: LUBANG: 1200 m, *Holttum SFN 25555* (K, SING); TENOMPOK: 1500 m, *Clemens 29419* (K).

28.6.17. Diplazium latisquamatum Holttum, Gardens' Bull. 9: 124 (1937).

Diplazium atrosquamosum sensu C. Chr. & Holttum p.p. non (Copel.) C. Chr., Gardens' Bull. 7: 274 (1934).

Terrestrial. Lower montane forest. Elevation: 1400–2700 m.

Material examined: KEMBURONGOH: 2700 m, *Clemens 27951* (K, SING); KIAU VIEW TRAIL: 1600 m, *Parris 10809* (K); KILEMBUN BASIN: 2600 m, *Clemens 33682* (SING); KILEMBUN RIVER: 1400 m, *Clemens 32516* (SING); LUMU-LUMU: 1800 m, *Clemens 28391* (K), 2100 m, *29716* (US); MARAI PARAI: 2100 m, *Clemens 32952* (K, SING); PENATARAN RIDGE: 1700 m, *Clemens 32454* (SING); PENATARAN RIVER: 2400 m, *Clemens 32551* (SING); TENOMPOK: 1500–1800 m, *Clemens 27122* (SING, US), 1700 m, *28103* (US), 1500 m, *28410* (K).

28.6.18. Diplazium lomariaceum (Christ) M. G. Price, Gardens' Bull. 36: 27 (1983).

Terrestrial. Lower montane forest. Elevation: 1400 m. Intergrades with *D. porphyrorachis*.

Material examined: KILEMBUN RIVER: 1400 m, *Clemens 34396* (K).

28.6.19. Diplazium megistophyllum (Copel.) Tagawa, Acta Phytotax. Geobot. 25: 180 (1973).

Athyrium megistophyllum Copel., Philipp. J. Sci. 56: 475, pl. 7 (1935). Type: Tinekuk Falls: 1500 m, *Clemens 40806* (holotype PNH?†; isotype K!). *Diplazium asperum* sensu C. Chr. & Holttum non Blume, Gardens' Bull. 7: 274 (1934). *Diplazium polypodioides* sensu C. Chr. & Holttum non Blume? (*Gibbs 4000*), Gardens' Bull. 7: 274 (1934).

Terrestrial. Lowland dipterocarp forest and lower montane forest. Elevation: 300–1500 m. Not endemic to Mount Kinabalu.

Additional material examined: KEBAYAU/KAUNG: 300 m, *Holttum SFN 25630* (K, SING); PARK HEADQUARTERS: 1500 m, *Parris 10834* (K); PORING HOT SPRINGS: 500 m, *Parris & Croxall 8994* (K); TENOMPOK: 1500 m, *Clemens 28080* (K).

28.6.20. Diplazium moultonii (Copel.) Tagawa, Acta Phytotax. Geobot. 25: 67 (1972).

Terrestrial. Tall upper montane forest. Elevation: 2700–3400 m.

Material examined: LUBANG/PAKA-PAKA CAVE: *Topping 1756* (SING, US); PAKA-PAKA CAVE: 3400 m, *Clemens 28975* (US), 2700 m, *Holttum SFN 25518* (K, SING); SUMMIT TRAIL: 2800 m, *Parris & Croxall 8835* (K).

28.6.21. Diplazium pallidum (Blume) T. Moore, Index Fil., 333 (1861). C. Chr. & Holttum, Gardens' Bull. 7: 269 (1934).

Terrestrial. Lowland dipterocarp forest to lower montane forest. Elevation: 600–1500 m.

Material examined: BAT CAVE: 600 m, *Parris & Croxall 8948* (K); LUBANG: 1200 m, *Holttum 25559* (K, SING); PINOSUK PLATEAU: 1500 m, *Parris & Croxall 9154* (K); TAKUTAN: 800 m, *Shea & Aban SAN 77222* (K).

28.6.22. Diplazium petiolare C. Presl, Epim., 86 (1849).

Terrestrial. Hill dipterocarp forest. Elevation: 1000 m.

Material examined: LOHAN/MAMUT COPPER MINE: 1000 m, *Beaman 10648* (K, MICH, MSC).

28.6.23. Diplazium poiense C. Chr. in C. Chr. & Holttum, Gardens' Bull. 7: 269 (1934).

Terrestrial. Lower montane oak forest. Elevation: 1400–1600 m.

Material examined: KIAU VIEW TRAIL: 1600 m, *Parris & Croxall 8545* (K); MEMPENING TRAIL: 1600 m, *Parris 11506* (K); TENOMPOK: 1400 m, *Holttum SFN 25380* (SING).

28.6.24. Diplazium porphyrorachis (Baker) Diels, Nat. Pfl. 1 (4): 225 (1899).

Asplenium porphyrorachis Baker, J. Bot. 17: 40 (1879). C. Chr. & Holttum, Gardens' Bull. 7: 279 (1934).

Terrestrial. Lower montane oak forest. Elevation: 1400–1700 m.

Material examined: LIWAGU RIVER TRAIL: 1700 m, *Parris 11472* (K), 1700 m, *Parris & Croxall 9103* (K); PARK HEADQUARTERS: 1500 m, *Parris & Croxall 8571* (K); TENOMPOK: 1500 m, *Clemens 28238* (K), 1400 m, *Holttum SFN 25379* (K, SING).

28.6.25. Diplazium sorzogonense (C. Presl) C. Presl, Tent., 114 (1836).

Diplazium sp. near japonicum (Thunb.) Bedd., C. Chr. & Holttum, Gardens' Bull. 7: 269 (1934).

Terrestrial. Hill forest. Elevation: 900 m.

Material examined: KADAMAIAN RIVER NEAR MINITINDUK: 900 m, *Holttum SFN 25579* (K).

28.6.26. Diplazium speciosum Blume, Enum. Pl. Javae, 193 (1828). C. Chr. & Holttum, Gardens' Bull. 7: 271 (1934).

Terrestrial. Lower montane valley oak forest. Elevation: 1500–2600 m.

Material examined: LAYANG-LAYANG: 2600 m, *Parris 11598* (K); PARK HEADQUARTERS: 1500 m, *Parris 11431* (K), 1500 m, *Parris & Croxall 8562* (K); SUMMIT TRAIL: 2500 m, *Parris 11546* (K), 2300 m, *Parris & Croxall 8664* (K), 2600 m, *8821* (K), 1800 m, *8885* (K); TENOMPOK/LUMU-LUMU: 1500 m, *Holttum SFN 25444* (SING).

28.6.27. Diplazium tabacinum Copel., Philipp. J. Sci. 1, Suppl.: 149, pl. 6 (1906). C. Chr. & Holttum, Gardens' Bull. 7: 274 (1934).

Terrestrial. Lower montane forest. Elevation: 1200–1500 m.

Material examined: LUBANG: 1200 m, *Holttum SFN 25565* (K, SING); TENOMPOK: 1500 m, *Clemens 26818* (K).

28.6.28. Diplazium tricholepis C. Chr. in C. Chr. & Holttum, Gardens' Bull. 7: 270, pl. 57 (1934). Type: Kemburongoh: 2100 m, *Holttum SFN 25522* (holotype BM n.v.; isotypes K!, SING!).

Terrestrial. Lower montane forest. Elevation: 1100–2100 m. Endemic to Mount Kinabalu.

Additional material examined: KIAU VIEW TRAIL: 1600 m, *Parris & Croxall 8546* (K); KILEMBUN BASIN: 1100 m, *Clemens 34476* (K, SING); LUMU-LUMU: 2100 m, *Clemens 29579* (K); PARK HEADQUARTERS: 1500 m, *Parris 11422* (K); TENOMPOK: 1500 m, *Clemens 29413* (K, SING).

28.6.29. Diplazium vestitum C. Presl

a. var. **borneense** C. Chr. in C. Chr. & Holttum, Gardens' Bull. 7: 273 (1934). Type: Dallas/Tenompok: 1400 m, *Clemens 27734* (syntype BM n.v.; isosyntype US!); Kaung/Dallas: 600 m, *Holttum SFN 25134* (syntype BM n.v.; isosyntypes K!, SING!); Kebayau/Kaung: 300 m, *Clemens 27676* (syntype BM n.v.; isosyntype K!); Tenompok: 1500 m, *Clemens 27478* (syntype BM n.v.; isosyntype K!), *Clemens 28690* (syntype BM n.v.).

Terrestrial. Lowland and lower montane valley oak forest. Elevation: 300–1600 m. Not endemic to Mount Kinabalu.

Additional material examined: LIWAGU RIVER TRAIL: 1600 m, *Parris 11483* (K), 1400 m, *Parris & Croxall 9037* (K).

28.6.30. Diplazium xiphophyllum (Baker) C. Chr., Index Fil., 241 (1905).

Asplenium xiphophyllum Baker, J. Bot. 17: 40 (1879). *Diplazium bantamense* Blume var. *alternifolium* (Blume) Alderw., Handb. Malayan Ferns, 406 (1908). C.

Chr. & Holttum, Gardens' Bull. 7: 274 (1934). *Diplazium fraxinifolium* C. Presl var. *grossum* (C. Presl) C. Chr. in C. Chr. & Holttum, Gardens' Bull. 7: 274 (1934).

Terrestrial. Hill forest and lower montane forest. Elevation: 900–1500 m.

Material examined: DALLAS: 1100 m, *Holttum SFN 25258* (K, SING); DALLAS/TENOMPOK: 1200 m, *Clemens 27733* (US); KUNDASANG: 1200 m, *RSNB 1438* (K, SING); TENOMPOK: 1500 m, *Clemens 27195* (SING), 1500 m, *29406* (K, SING), 1500 m, *29438* (K).

28.6.31. Diplazium sp. 1

Terrestrial. Lower montane valley forest. Elevation: 2600 m. Endemic to Mount Kinabalu.

Material examined: SUMMIT TRAIL: 2600 m, *Parris & Croxall 8813* (K).

Acknowledgments

We especially appreciate the permission to work in Kinabalu Park by the Director of Sabah Parks, Lamri Ali. In 1988 the Director General, Socio-economic Research Unit, Prime Minister's Department, Kuala Lumpur, generously gave permission for Parris to work in the Kinabalu and Crocker Range Parks. The logistic support provided by Director Lamri and his able staff, the Park Ecologist Jamili Nais and former Ecologists Peter Walpole and Anthea Phillipps, Naturalist Answ Gunsalam, and Rangers Gabriel Sinit and Justin Jukien, all helped in many ways to make our visits successful.

The decision of Parris to document the fern flora of Mount Kinabalu owes a great debt to Professor E. J. H. Corner, who not only inspired her to go and see it for herself but was also a great support in fund-raising for her first visit in 1980. Funds for this fieldwork were provided by the Eileen and Phyllis Gibbs Travelling Fellowship 1980–81 from Newnham College, Cambridge, and the Percy Sladen Memorial Fund. She is grateful to both for enabling her to prepare the foundations of this study. On this first visit Dr. J. P. Croxall provided extensive field assistance, particularly in the collecting of Hymenophyllaceae, in breaks from bird-watching. Expedition funds from the Royal Botanic Gardens, Kew, enabled further collecting to be carried out, and Parris thanks Brian Spooner and Susyn Andrews of Kew for their company in 1985 and 1988, respectively.

A Fulbright Fellowship sponsored the field research of the Beaman team in 1983-84. Subsequent research has been supported by NSF grants BSR-8507843 and BSR-8822696 to Michigan State University.

Field work of the Beamans was aided by many persons in 1983–84 who came for their own studies but served effectively as participants in collecting pteridophytes and many other plants. Teofila Beaman played a particularly important role in helping our guests, providing keen eyesight in the field, and spending many long evenings and days pressing and sorting plants. Dr. Ghazally Ismail, formerly Dean of the Faculty of Science and Natural Resources of the National University of Malaysia, Sabah Campus, facilitated in many ways the field work of the Beamans. Jacinto C. Regalado, Jr. participated extensively in preliminary phases of producing the enumeration, including computer programming and recording much of the pteridophyte specimen data at Kew and Singapore. Michael G. Price identified all the Beaman fern collections and has assisted in many other ways, particularly in enabling our use of the Copeland Herbarium at the University of Michigan and in providing much taxonomic advice. Benito Tan determined some of the Beaman collections of Lycopodiaceae and *Selaginella*. A. F. Clewell provided support and facilities for the participation of Reed Beaman. Robert Johns and Peter Edwards of the Fern Section at Kew and Peter Hovenkamp of the Rijksherbarium, Leiden, likewise provided able assistance in our access to the specimens and library resources under their supervision.

Publication of this enumeration as one of the first database products coming directly in camera-ready copy from a computer has been made possible through support of the Royal Botanic Gardens, Kew, particularly with the encouragement of the Keeper of the Herbarium, G. Ll. Lucas. Much help also has been provided by the Head of the Computer Section, W. Loader, Editor of the Kew Bulletin, J. M. Lock, the former Editor of the Kew Bulletin, M. J. E. Coode, and R. K. Brummitt, P. J. Cribb, J. Dransfield, R. Polhill, and C. E. Powell.

We appreciate the opportunity to use the facilities or the loan of specimens from the following herbaria: BM, BO, CGE, K, KYO, L, LAE, MICH, MSC, NY, SAN, SING, UC, US. It should be noted, however, that the only herbaria in which an attempt was made to examine all relevant specimens were K and SING. AK provided facilities for loans for Parris.

Literature Cited

Beaman, J. H. & Beaman, R. S. 1990. Diversity and distribution patterns in the flora of Mount Kinabalu. *In* The Plant Diversity of Malesia, P. Baas, K. Kalkman, & R. Geesink, eds. Kluwer Academic Publishers, Dordrecht/Boston/London, pp. 147–160.

Beaman, J. H. & Regalado, J. C., Jr. 1989. Development and management of a microcomputer specimen-oriented database for the flora of Mount Kinabalu. Taxon 38: 27–42.

Christensen, C. & Holttum, R. E. 1934. The ferns of Mount Kinabalu. Gardens' Bull. 7: 191–324.

Gibbs, L. S. 1914. A contribution to the flora and plant formations of Mount Kinabalu and the highlands of British North Borneo. J. Linn. Soc., Bot. 42: 1–240, 8 pl.

Holttum, R. E. 1978. The ferns of Kinabalu National Park. Chapter 8. *In* Kinabalu: Summit of Borneo. M. Luping, Chin Wen & E. R. Dingley, eds. Sabah Society Monograph 1978. Sabah Society, Kota Kinabalu, Sabah, pp. 199–210.

Jacobsen, W. B. G., & Jacobsen, N. H. G. 1989. Comparison of the pteridophyte floras of Southern and Eastern Africa, with special reference to high-altitude species. Bull. Jard. Bot. Nat. Belg. 59: 261–317.

Johns, R. J. & Stevens, P. F. 1971. Mount Wilhelm Flora: A check list of the species. Botany Bull. 6, Div. of Botany, Dept. of Forests, Lae, New Guinea, 60 pp., 1 map.

Parris, B. S. 1985. Ecological aspects of distribution and speciation in Old World tropical ferns. Proc. Roy. Soc. Edinb. 86B: 341–346.

Parris, B. S., Jermy, A. C., Camus, J. M. & Paul, A. M. 1984. The Pteridophyta of Gunung Mulu National Park. *In* Studies on the Flora of Gunung Mulu National Park, Sarawak. A. C. Jermy, ed. Forest Department, Sarawak, Kuching, pp. 145–233.

Stapf, O. 1894. On the flora of Mount Kinabalu, in North Borneo. Trans. Linn. Soc. London, Bot. 4: 69–263, pl. 11–20.

INDEX TO NUMBERED COLLECTIONS CITED

Aban 79577 (2.3.2).

Abbe 10196 (16.15.1).

Abbe et al. 9930 (2.1.2); 9955 (3.1.1); 9968 (15.3.17); 9975 (23.9.1); 9981 (9.4.3); 9990 (10.4.17a); 9998 (10.2.2).

Beaman 6772 (2.3.1); 7189 (2.1.2); 7456 (6.1.21); 7460 (13.13.1); 7476 (2.1.7); 7478 (28.6.7a); 7493 (23.16.3a); 7561 (21.2.1); 7575 (26.11.6); 7978 (2.1.5); 7979 (15.4.4); 7980 (10.4.3); 7996 (10.9.3); 8121 (10.4.17c); 8135 (2.1.7); 8144 (24.1.13); 8145 (14.3.4); 8210 (2.1.5); 8364 (23.20.4); 8439 (10.4.15); 8498 (23.2.2); 8555 (23.2.2); 8578 (27.1.6); 8586 (10.4.17c); 8610 (5.7.1); 8625 (8.1.3); 8626 (23.11.5); 8649 (15.3.15); 8651 (23.2.2); 8652 (23.13.11); 8662 (15.4.4); 8663 (23.10.1); 8664 (6.1.22); 8666 (2.1.5); 8669 (6.1.6); 8675 (9.1.1); 8677 (15.5.2); 8685 (6.1.6); 8697 (15.4.20); 8746 (8.1.16); 8772 (15.4.20); 8780 (27.1.6); 8784 (11.1.1); 8787 (12.2.1); 8801 (13.16.5); 8803 (2.1.5); 8872 (23.13.6); 8873 (23.14.1); 8986 (5.7.1); 8987 (25.2.1); 9003 (6.1.6); 9004 (6.1.22); 9026 (2.1.7); 9027 (2.1.5); 9028 (10.4.3); 9030 (27.1.6); 9033 (23.16.3a); 9050 (2.1.6); 9054 (5.1.3); 9055 (10.4.7); 9058 (23.20.3); 9059 (13.16.11); 9067 (25.2.1); 9100 (5.4.1); 9105 (6.1.18); 9106 (15.4.4); 9112 (17.2.2); 9116 (2.1.7); 9117 (2.1.5); 9130 (24.1.13); 9204 (20.2.4); 9237 (23.6.1); 9243 (23.20.4); 9286 (13.16.10); 9295 (23.14.1); 9304 (23.12.1); 9320 (23.14.1); 9321 (23.2.2); 9322 (27.1.8); 9357 (27.1.7); 9358 (23.11.5); 9359 (23.13.11); 9372 (23.13.5); 9373 (23.2.2); 9536 (6.1.6); 9538 (6.1.5); 9539 (7.1.11); 9540 (12.2.1); 9541 (26.13.7); 9564 (23.2.2); 9566 (15.4.4); 9590 (9.4.2); 9605 (20.3.1); 9608 (6.1.10); 9609 (16.7.2); 9613 (15.4.7); 9613a (15.4.22); 9806 (23.13.9); 9807 (23.20.5); 9808 (6.1.21); 9809 (6.1.5); 9810 (15.5.1); 9836 (13.13.4); 9867 (23.4.5); 9872 (2.1.2); 9873 (2.1.4); 9885 (25.2.3); 9921 (20.3.1); 9926 (10.4.3); 9927 (23.4.5); 9942 (2.1.2); 9943 (15.3.12); 9943a (15.5.2); 9947 (26.4.2a); 9955 (10.10.3a); 9964 (2.3.2); 10330 (12.1.1); 10332 (20.3.1); 10522 (9.4.2); 10523 (23.16.3a); 10524 (23.18.1); 10527 (23.2.2); 10528 (16.2.2); 10529 (6.1.34); 10543 (15.5.1); 10546 (15.2.3); 10548 (28.6.7a); 10557 (16.3.3); 10569 (27.3.4); 10602 (23.13.4); 10603 (23.6.1); 10624 (23.20.3); 10624a (9.4.1); 10625 (23.20.5); 10640 (26.11.5); 10641 (28.6.10); 10648 (28.6.22); 10649 (26.13.4); 10658 (23.19.1); 10662 (2.1.4); 10663 (6.1.22); 10678 (2.1.7); 10679 (6.1.6); 10724 (28.6.5); 10725 (26.13.4); 10726 (10.4.2); 10727 (26.11.5); 10728 (16.2.2); 10729 (7.1.2); 10730 (5.7.1); 10731 (23.13.6); 10732 (23.16.4); 10733 (23.4.2); 10733a (5.7.1); 10734 (13.4.2); 10735 (23.4.2); 10736 (23.2.2); 10751 (23.4.2); 10755 (10.10.3a); 10775 (28.6.12); 10776 (9.1.1); 10777 (16.6.1); 10779 (15.4.4); 10781 (23.16.1); 10784 (6.1.22); 10787 (17.2.3); 10789 (2.1.7); 10796 (2.1.6); 10799 (23.10.1); 10908 (1.1.1a); 10991 (10.4.14).

Carr 26830 (26.11.6).

Chai & Ilias 6005 (6.1.9).

Chew & Corner 4088 (21.3.2b); 4120 (6.1.16); 4128 (18.1.1); 4301 (2.1.5); 4331 (8.1.16); 4413 (2.3.2); 4414 (23.16.4); 4416 (20.2.2); 4582 (2.1.7); 4617 (17.2.6); 4710 (10.4.14); 4715 (11.4.1); 4716 (8.1.3); 4717 (13.3.2); 4785 (10.9.3); 4988 (21.1.1); 5981 (2.1.1); 7139 (2.3.3); 7143 (14.2.2).

Chew, Corner & Stainton 12 (23.10.1); 38 (17.4.1); 123 (27.1.1); 147 (5.4.1); 170 (8.1.3); 175 (16.6.1); 180 (5.4.1); 181 (23.4.5); 183 (23.4.1); 187 (2.3.1); 203 (8.1.8); 260 (20.2.3); 298 (20.3.1); 300 (13.13.1); 576 (26.13.6); 577 (23.13.4); 579 (13.12.1); 580 (13.16.1); 588 (12.1.1); 590 (5.7.1); 593 (2.1.5); 636 (23.22.1); 667 (13.7.1); 673 (16.8.6); 722 (15.2.3); 723 (15.3.8); 725 (23.2.2); 726 (2.3.4); 727 (2.1.1); 729 (22.1.2); 730 (26.4.2a); 735 (13.1.1); 772 (15.4.8); 773 (16.7.3); 774 (9.4.4); 786 (10.2.1); 790 (8.1.9); 791 (6.1.10); 794 (28.6.13); 794b (23.2.3a); 795 (5.4.1); 799 (15.3.21); 801 (23.11.4); 803 (16.7.2); 804 (23.2.3a); 805 (23.16.2); 829 (15.4.22); 834 (13.9.4); 836 (15.4.6); 849 (15.3.20); 850 (16.10.3); 851 (15.4.13); 854 (23.4.4); 879 (23.16.2); 882 (16.7.2); 883 (28.2.2); 884 (7.1.4); 905 (14.2.2); 907 (16.12.2); 908 (2.3.5); 911 (2.3.1); 915 (17.2.1); 918 (13.9.8); 921 (2.1.1); 923 (10.3.1); 926 (14.2.1); 946 (24.1.16); 947 (13.6.1); 948 (16.8.5); 949a (8.1.9); 949b (13.13.5); 950 (24.1.10); 951 (16.14.1); 952 (10.3.5); 990 (8.1.15); 995 (14.3.3a); 998 (11.1.1); 1022 (10.10.3a); 1140 (23.4.5); 1141 (16.12.2); 1142 (16.8.6); 1143 (26.4.2a); 1144 (15.3.4); 1145 (15.2.3); 1146a (15.4.6); 1146b (15.4.14); 1147 (15.3.21); 1148 (15.4.8); 1149 (15.5.2); 1150 (16.7.2); 1151 (23.16.2, 23.16.3a); 1152 (15.3.12); 1153 (15.4.8); 1154 (26.4.2a); 1155 (13.1.1); 1155A (13.9.4); 1155b (13.1.1); 1156 (10.2.1); 1157 (23.4.4); 1158 (9.4.4); 1159 (23.4.2); 1159b (15.4.22); 1159c (15.4.14); 1160 (15.6.4); 1164 (23.16.3a); 1165 (23.16.4); 1172 (26.13.6); 1183 (23.13.4); 1186 (13.16.4); 1212 (13.4.3a); 1213 (26.13.9); 1214 (26.10.4); 1216 (26.2.1); 1228 (23.11.1); 1231 (21.3.3a); 1232 (28.4.1); 1240 (23.4.1); 1245 (21.3.3a); 1257 (21.3.3a); 1270 (23.4.5); 1273 (20.3.1); 1294 (23.10.1); 1295 (6.1.16); 1296 (20.2.2); 1303 (10.9.1a); 1305 (23.16.4); 1310 (26.13.7); 1341 (7.1.1); 1354 (6.1.31); 1362 (15.3.13);

1367 (6.1.20); 1368 (9.1.1); 1373 (26.10.1); 1374 (10.4.14); 1377 (23.11.3); 1389 (23.13.8); 1402 (23.13.9); 1404 (23.20.3); 1420 (23.3.3); 1420B (23.13.11); 1420a (23.3.3); 1420b (23.13.11); 1432 (23.3.2); 1438 (28.6.30); 1440 (15.5.2); 1451 (10.4.12); 1453 (10.4.15); 1461 (20.3.1); 1464 (10.4.17c); 1501 (16.16.2); 1505 (8.1.8); 1526 (8.1.16); 1529 (10.10.3a); 1532 (18.3.1); 1533 (28.5.1); 1568 (5.7.2); 1578 (10.6.1); 1580a (10.4.15); 1580b (10.4.4); 1581 (7.1.1); 1582 (7.1.3); 1593 (16.10.2); 1594 (16.8.6); 1597 (16.16.1); 1598 (16.3.3); 1651 (6.1.9); 1666 (16.8.5); 1677 (10.1.3); 1678 (26.10.1); 1679 (26.2.1); 1709 (27.1.7); 1720 (23.13.9); 1721 (6.1.31); 1727 (23.20.1); 1736 (16.8.5); 1737 (6.1.21); 1739 (16.7.9); 1741 (2.1.4); 1742 (10.4.17c); 1743 (16.7.5); 1786 (23.19.2); 1811 (10.4.18); 1814 (2.1.7); 1869 (15.2.7, 15.3.22); 1886 (8.1.10); 1887 (2.1.3); 1889 (2.1.6); 2005 (16.10.4); 2630 (13.13.2); 2731 (10.2.2); 2765 (24.1.12); 2812 (18.1.4); 2853 (13.12.1); 2854 (8.1.8); 2859 (23.22.1); 2896 (10.4.17c); 2897 (6.1.28); 2898 (8.1.21); 2899 (11.4.1); 2975 (10.4.6); 2979 (8.1.10).

Clemens 9986 (14.2.4); 10093 (21.3.3a); 10226 (16.8.12); 10231 (10.4.17c); 10234 (18.1.5); 10241 (13.12.3); 10322 (9.3.1); 10323 (24.1.15); 10324 (6.1.19a); 10325 (27.3.7); 10327 (5.1.1); 10347 (26.10.1); 10348 (24.1.2); 10350 (24.1.12); 10355 (13.6.1); 10356 (9.2.2); 10358 (5.1.1); 10388 (9.1.1); 10443 (1.1.1a); 10446 (8.1.4); 10500 (8.1.14); 10520 (2.3.1); 10530 (15.4.7); 10533 (14.3.4); 10587 (23.16.2); 10588 (16.5.1); 10590 (22.1.4); 10613 (2.1.1); 10618p.p. (15.4.3); 10620 (13.15.1); 10621 (28.2.3); 10648 (13.13.3); 10649 (15.4.13); 10717 (15.4.6); 10721 (15.7.3); 10722 (15.6.4); 10724 (14.2.2); 10726 (8.1.7); 10728 (26.13.8); 10729 (25.2.3); 10780 (16.8.12); 10855 (15.3.7); 10859 (8.1.20); 10860 (28.6.6); 10909 (10.4.9); 10914 (28.6.8); 10915 (8.1.10); 10916 (6.1.5); 10919 (25.2.4a); 10920 (15.4.22); 10921 (8.1.6); 10947 (14.3.2); 11011 (28.4.1); 11032 (7.1.5); 11033 (8.1.3); 11034 (15.5.2); 11040 (15.3.20); 11043 (15.2.1); 11044 (15.2.8); 11047 (15.4.19); 11048 (15.6.4); 11049 (15.3.12); 11051 (28.6.3); 11053 (20.3.1); 11056 (16.6.1); 11058 (15.3.10); 11063 (10.4.9); 11064 (15.4.14); 11068 (26.4.2a); 11069 (26.4.1); 11072 (15.7.4); 26005 (23.15.1); 26145 (9.2.2); 26158 (21.3.3a); 26184 (23.3.2); 26384 (27.1.1); 26404 (2.1.2); 26488 (23.22.1); 26489 (23.11.1); 26495 (23.17.1); 26646 (6.1.16); 26650 (4.1.7); 26651 (10.4.17a); 26704 (23.10.1); 26766 (26.13.10a); 26800 (23.13.3); 26809 (10.3.3); 26818 (28.6.27); 26818b (6.1.33); 26824 (8.1.12); 26825 (18.1.2); 26826 (18.1.2); 26827 (10.1.3); 26830 (6.1.28); 26845 (13.16.7); 26846 (13.4.3a); 26850 (16.8.6); 26852 (11.1.1); 26855 (23.8.3); 26858 (16.3.1); 26859 (16.13.5); 26861 (8.1.4); 26889 (6.1.16); 26890 (6.1.21); 26892 (23.22.1); 26893 (26.13.5); 26894 (6.1.11); 26895 (10.4.2); 26897 (16.16.2); 26898 (28.6.7a); 26908 (26.13.12); 26923 (26.11.3); 26924 (6.1.1); 26925 (20.2.3); 26928 (26.13.16a); 26929 (23.13.11); 26931 (13.16.4); 26932 (26.13.1); 26942 (23.13.9); 26948 (16.7.2); 26949 (23.11.3); 26950 (28.6.7a); 26975 (23.2.1, 23.2.2); 26976 (10.1.2); 26978 (10.5.3); 26980 (10.3.4); 26981 (15.5.2); 26982 (5.1.1); 26983 (26.13.5); 26984 (23.20.2); 27007 (28.2.3, 9.4.3); 27020 (4.1.9); 27025 (23.13.5); 27029 (26.10.4); 27036 (5.5.1); 27037 (23.11.1); 27043 (22.1.4); 27046 (6.1.10); 27046b (6.1.13); 27047 (7.1.4); 27048 (23.2.3a); 27049 (28.2.3); 27050 (15.4.6); 27057 (6.1.10); 27059 (15.4.6); 27060 (17.2.8); 27061 (10.4.18); 27062 (28.6.7a); 27063 (23.4.5); 27065 (15.4.22); 27065A (15.2.1); 27066 (28.6.7b); 27068 (24.1.1); 27070 (16.8.11); 27071 (16.6.1); 27088 (2.1.1); 27118 (24.1.5); 27121 (9.1.1); 27122 (28.6.17); 27137 (27.1.1, 26.10.2); 27138 (16.13.2); 27144 (17.5.1); 27161 (23.4.5); 27182 (28.2.3); 27189 (10.4.17c); 27190 (26.13.2); 27195 (28.6.30); 27208 (4.1.8); 27210 (20.3.1); 27213 (24.1.1); 27218 (5.7.1); 27227 (5.3.2); 27229 (24.1.7); 27231 (6.1.35); 27267 (10.4.5); 27288 (26.13.10a); 27289 (24.1.9); 27290 (23.22.1); 27290b (23.22.1); 27292 (6.1.11); 27293 (23.20.3); 27294 (23.13.5); 27295 (13.16.10); 27296 (23.5.1); 27297 (20.2.6); 27298 (6.1.30); 27300 (23.11.1); 27304 (6.1.26a); 27305 (23.13.3); 27307 (23.11.2); 27309 (6.1.35); 27313 (13.16.8); 27315 (16.7.1); 27317 (26.13.10a); 27327 (23.20.5); 27328 (13.3.2); 27329 (13.16.7); 27331 (8.1.12); 27349 (23.13.9); 27350 (23.8.1); 27353 (20.2.3); 27354 (18.1.2); 27357 (25.2.2); 27373 (6.1.28); 27383 (20.1.1); 27384 (13.14.1); 27385 (28.6.7a); 27402 (27.1.4); 27407 (10.5.1); 27420 (1.1.1a); 27423 (23.4.2); 27425 (16.8.1); 27428 (18.3.1); 27430 (26.11.3); 27431 (23.8.4); 27433 (28.6.7b); 27441 (27.1.2); 27443 (27.3.3); 27454 (27.1.1); 27455 (27.1.7); 27458 (26.13.12); 27465 (10.1.3); 27468 (23.3.2b); 27469 (26.3.2); 27470 (14.3.3a); 27471 (26.13.2); 27475 (26.3.2); 27477 (6.1.28); 27478 (28.6.29a); 27479 (23.20.2); 27479b (23.20.2); 27485 (24.1.4); 27492 (27.1.1); 27502 (28.6.7a); 27503 (23.2.2); 27512 (27.3.7); 27513 (5.1.1); 27514 (4.1.1); 27523 (26.1.1); 27526 (26.6.1); 27527 (10.5.2); 27541 (26.10.4); 27543 (26.13.9); 27564 (23.16.4); 27578 (13.12.3); 27588 (16.3.2); 27596 (17.4.1); 27600 (4.1.10a); 27601 (26.13.1); 27603 (26.13.5); 27604 (6.1.16); 27653 (17.1.2); 27660 (13.3.2); 27663 (26.11.4); 27666 (21.2.1); 27672 (10.5.4a); 27676 (28.6.29a); 27677 (11.3.1); 27678 (5.2.1); 27688 (6.1.36); 27690 (24.1.4); 27729 (26.13.9); 27730 (24.1.12); 27733 (28.6.30); 27734 (28.6.29a); 27736 (16.16.1); 27737 (6.1.9); 27739 (26.1.1, 26.3.2); 27740 (11.1.1); 27741 (8.1.19); 27742 (11.2.1); 27744 (10.4.4); 27746 (6.1.31); 27753 (13.1.1); 27755 (7.1.12); 27756 (15.3.8); 27758 (2.1.5); 27763 (26.11.4); 27769 (6.1.1); 27775 (2.1.1); 27782 (28.2.3); 27783 (15.4.13); 27789 (10.3.3); 27790 (9.2.2); 27791 (16.7.2); 27792 (15.4.4, 15.4.20); 27803 (9.2.2); 27816 (9.4.3); 27839 (10.10.3a); 27892 (21.1.1); 27896 (15.4.6); 27904 (23.16.1); 27907 (10.5.5); 27909 (24.1.2); 27925 (16.7.2); 27933 (23.16.4); 27937 (8.1.21); 27944 (28.2.3); 27946 (13.15.1); 27947 (28.2.2); 27948 (13.13.2); 27950 (10.5.1); 27951 (28.6.17); 27952 (25.2.3); 27953 (27.3.1); 27954 (8.1.14); 27955 (23.2.3a); 27956 (8.1.7); 27957 (10.7.1); 27959 (8.1.3); 27960 (22.1.1); 27962 (23.16.2); 27963 (6.1.22); 27964 (6.1.10); 27965 (15.3.6); 27966 (13.8.1); 27968 (13.13.3); 27969 (13.3.3); 27970 (26.4.3); 27971 (26.4.2a); 27973 (13.13.2); 27976 (13.9.3); 27988 (3.1.1); 27989 (15.7.3); 27990 (15.4.13); 27991 (15.4.14, 16.7.2); 27992 (15.1.1); 27993 (15.5.3); 27994 (15.3.12);

27995 (15.4.8); 27997 (15.4.17); 27998 (15.4.1, 15.4.29, 15.6.2b); 27999 (15.4.6); 28000 (15.2.1); 28001 (10.4.18); 28005 (16.8.2); 28006 (16.8.9); 28007 (16.12.1); 28010 (16.7.2); 28014 (16.7.4); 28015 (14.3.4); 28016 (14.4.1a); 28018 (10.2.1); 28019 (17.2.2, 17.2.8); 28020 (17.2.1); 28032 (13.3.2); 28038 (6.1.1); 28052 (6.1.16); 28054 (6.1.16); 28065 (20.2.3); 28080 (28.6.19); 28082 (26.13.9); 28084 (16.8.5); 28085 (15.2.10); 28088 (16.8.2); 28093 (24.1.12); 28094 (13.13.1); 28095 (24.1.2); 28097 (9.1.1); 28098 (26.4.3); 28099 (26.13.5); 28102 (13.3.2); 28103 (28.6.17); 28104 (16.7.4); 28153 (23.16.1); 28154 (4.1.3); 28155 (28.6.8); 28157 (15.3.7); 28158 (8.1.8); 28165 (6.1.21); 28172 (24.1.3); 28175 (10.5.2); 28185 (27.3.3); 28191 (23.2.1); 28199 (23.10.1); 28205 (13.9.5); 28207 (16.10.4); 28215 (10.4.13); 28225 (16.1.1, 16.14.1); 28228 (16.7.6); 28232 (16.3.2); 28233 (26.13.3); 28236 (10.5.2); 28238 (28.6.24); 28240 (15.4.22); 28245 (23.16.3a); 28266 (8.1.16); 28267 (14.2.4); 28268 (10.2.2); 28281 (10.1.1); 28282 (26.13.3); 28287 (13.12.1); 28302 (16.3.2); 28315 (15.2.1); 28317 (15.5.2); 28320 (21.3.2a); 28337 (20.3.1); 28338 (28.6.7b); 28339 (12.1.1); 28341 (10.1.1); 28344 (23.18.1); 28350 (10.10.3a); 28360 (5.1.1); 28362 (10.2.2); 28364 (6.1.5); 28366 (10.3.3); 28372 (16.16.2); 28382 (26.13.3); 28385 (22.1.3); 28386 (16.8.5); 28387 (23.4.1, 23.13.5); 28388 (16.7.2); 28391 (28.6.17); 28397 (16.8.5); 28410 (28.6.17); 28412 (28.6.11); 28421 (16.3.2); 28441 (15.3.14, 15.3.21); 28446 (9.1.1); 28453 (26.13.1); 28456 (26.13.2); 28457 (6.1.28); 28459 (16.14.1); 28460 (19.1.1); 28532 (2.3.2); 28544 (23.18.1); 28545 (9.1.1); 28546 (6.1.21); 28550 (10.5.2); 28577 (6.1.8); 28596 (8.1.16); 28597 (10.4.12); 28598 (13.9.2); 28603 (10.4.14); 28638 (17.3.1); 28641 (6.1.10); 28642 (23.3.2); 28645 (10.4.4); 28646 (26.10.4); 28681 (15.5.2); 28684 (23.6.2); 28685 (23.20.3); 28689 (6.1.1); 28690 (28.6.29a); 28695 (8.1.21); 28727 (7.1.4); 28728 (7.1.3); 28734 (11.4.1); 28738 (2.2.1); 28746 (2.1.1); 28749 (13.12.3); 28763 (20.3.1); 28785 (14.4.1a); 28826 (10.4.17c); 28837 (13.1.1); 28850 (8.1.15); 28852 (13.12.3); 28857 (23.16.2); 28861 (2.3.6); 28879 (6.1.21); 28883 (7.1.12); 28960 (22.1.4); 28961 (22.1.2); 28962 (22.1.2); 28963 (10.4.18); 28963b (10.4.18); 28964 (26.4.2a); 28966 (15.3.6); 28966b (15.3.6); 28967 (15.3.12); 28968 (15.3.4); 28968A (15.4.20); 28969 (15.3.12); 28971 (13.13.2); 28973 (13.8.1); 28974 (10.7.1); 28975 (28.6.20); 28977 (13.1.1); 28978 (15.3.14); 28980 (2.3.6); 28981 (15.4.13); 28981A (15.4.17); 28982 (15.6.4); 28983 (16.15.1); 28984 (16.8.11); 28985 (16.6.1); 28987 (15.4.6); 28992 (22.1.2); 29003 (2.3.2); 29011 (23.12.1); 29028 (24.1.4); 29029 (15.3.2); 29030 (23.4.2); 29033 (28.2.2); 29035 (15.4.14); 29038 (23.1.1); 29044 (15.3.5, 15.3.13); 29045 (28.2.1); 29046 (13.6.1); 29047 (10.7.1); 29048 (4.1.11); 29050 (26.13.3); 29052 (28.4.1); 29053 (13.13.5); 29054 (26.13.2); 29055 (11.4.1); 29055A (11.4.1); 29056 (6.1.21); 29057 (13.9.6); 29058 (15.2.2); 29059 (22.1.2); 29060 (22.1.3); 29061 (15.4.4); 29062 (26.2.1); 29063 (9.4.4); 29064 (17.2.6); 29065 (2.1.5); 29066 (26.4.2a); 29067 (16.16.2); 29068 (8.1.11); 29092 (10.5.1); 29104 (15.6.4); 29104A (15.2.1); 29105 (9.2.4); 29106 (23.2.1, 23.2.2); 29107 (10.8.1); 29109 (8.1.3); 29110 (10.3.4); 29111 (13.12.3); 29161 (10.9.1a); 29162 (11.1.1); 29210 (21.1.1); 29228 (6.1.22); 29229 (23.11.2); 29231 (23.13.11); 29238 (26.3.2); 29239 (10.3.4); 29240 (26.13.16a); 29241 (20.2.5); 29242 (26.5.2); 29266 (6.1.12); 29270 (16.8.12); 29275 (28.6.11); 29276 (13.1.1); 29277 (23.4.5); 29277b (23.4.5); 29280 (21.3.3a); 29299 (23.20.3); 29305 (6.1.9); 29311 (24.1.2); 29312 (23.3.1); 29332 (17.2.5); 29333 (26.4.2a); 29334 (12.1.1); 29347 (16.16.1); 29395 (8.1.21); 29405 (2.3.4); 29406 (28.6.30); 29411 (13.16.4); 29413 (28.6.28); 29416 (13.9.5); 29417 (27.1.7); 29418 (6.1.1); 29419 (28.6.16); 29422 (28.6.7a); 29429 (6.1.33); 29430 (14.2.4); 29435 (10.1.1); 29438 (28.6.30); 29450 (26.10.4); 29462 (5.1.1); 29484 (23.8.3); 29485 (10.3.3); 29488 (16.7.2); 29491 (8.1.15); 29492 (7.1.3); 29506 (15.1.1, 5.1.1); 29507 (10.4.17c); 29516 (13.16.8); 29523 (26.1.1); 29535 (14.1.3a); 29545 (23.4.3); 29548 (20.2.4); 29562 (9.3.1); 29566 (17.5.1); 29567 (24.1.8); 29568 (20.1.1); 29569 (8.1.19); 29573 (17.5.1); 29574 (27.2.2); 29579 (28.6.28); 29581 (27.3.7); 29582 (23.13.1); 29602 (10.1.1); 29618 (28.6.9); 29623 (4.1.5); 29624 (23.3.1); 29625 (23.8.4); 29626 (9.4.2); 29627 (23.8.4), 29628 (23.20.3); 29629 (13.12.3, 13.13.4); 29668 (23.7.1); 29672 (11.1.1); 29683 (10.5.2); 29698 (2.1.7); 29705 (23.4.2); 29709 (17.2.6); 29710 (13.9.1); 29711 (6.1.10); 29712 (8.1.9); 29714 (8.1.11); 29715 (16.7.2); 29716 (28.6.17); 29734 (11.4.1); 29749 (28.6.13); 29765 (26.1.1); 29765a (26.1.2); 29766 (28.6.3); 29805 (8.1.7); 29830 (6.1.13); 29875 (13.9.4); 29912 (15.3.5); 29937 (13.13.4); 29938 (5.4.1); 29940 (13.9.6); 29941 (10.5.2); 29966 (13.1.1); 29967 (28.2.1, 5.4.1); 29968 (23.3.1); 29969 (11.4.1); 29970 (6.1.31); 29971 (23.16.3a); 29972 (6.1.20); 29973 (24.1.8); 29982 (9.5.1); 29987 (8.1.3); 30378 (6.1.32); 30446 (28.2.4); 30447 (15.3.9); 30455 (13.7.1); 30456 (26.13.1); 30457 (17.4.1); 30458 (26.11.1); 30459 (18.3.1); 30460 (10.5.1); 30461 (15.2.3); 30462 (15.4.6); 30463 (26.10.2); 30480 (23.13.1); 30490 (10.5.1); 30523 (15.4.6); 30552 (13.8.2); 30553 (26.13.15); 30609 (15.7.1); 30611 (27.1.7); 30615 (6.1.34); 30622 (24.1.3); 30636 (17.3.1); 30668 (13.7.1); 30676 (13.2.1); 30702 (10.3.1); 30703 (23.13.9); 30704 (23.8.3); 30705 (20.3.1); 30733 (10.4.9); 30735 (8.1.10); 30736 (26.13.15); 30737 (7.1.2); 30741 (16.14.1); 30748 (15.3.2); 30756 (17.5.1); 30809 (17.5.1); 30841 (5.1.2); 30871 (17.5.1); 30890 (17.5.1); 30902 (20.3.1); 30903 (10.4.18); 30906 (8.1.18); 30933 (10.4.1); 30954 (9.3.2); 30962 (8.1.18); 30963 (8.1.10); 30985 (15.1.1); 31029 (13.16.5); 31038 (13.16.5); 31039 (23.8.2); 31059 (27.3.1); 31084 (6.1.31); 31086 (24.1.5); 31088 (10.4.18); 31145 (16.8.11); 31175 (16.14.1); 31251 (16.14.1); 31305 (5.7.1); 31309 (6.1.5); 31346 (17.5.1); 31365 (23.19.2); 31390 (8.1.15); 31392 (14.4.1a); 31396 (13.6.1); 31415 (16.12.2); 31423 (15.2.1); 31423A (15.2.3); 31435 (7.1.2); 31469 (6.1.28); 31610 (17.5.1); 31698 (8.1.5); 31706 (17.5.1); 31725 (7.1.11); 31729 (15.4.4); 31740 (23.4.5); 31743 (11.4.1); 31770 (26.4.1); 31773 (7.1.11); 31783 (16.7.2); 31784 (15.3.20); 31787 (23.16.3a); 31789 (15.3.19); 31794 (15.3.21); 31798 (28.2.2); 31799 (17.2.7); 31804 (16.8.5); 31822 (22.1.2); 31869 (17.2.7); 31871 (23.16.2); 31921 (15.4.14); 31922 (13.13.5); 31923 (8.1.15); 31953 (13.3.3); 31955 (13.9.4, 13.9.7); 31957 (7.1.5); 31958 (9.4.4); 31959 (28.2.2); 31960

(16.10.1); 31989 (22.1.2); 31992 (11.4.1); 31994 (14.2.1); 32018 (16.8.1); 32033 (23.13.2); 32058 (9.4.4); 32065 (13.16.7); 32066 (18.3.1); 32067 (6.1.32); 32122 (23.1.1); 32142 (15.4.9); 32184 (2.1.6); 32250 (16.8.6); 32287 (15.4.23); 32298 (8.1.3); 32299 (26.11.7a); 32328 (10.4.9); 32347 (15.4.13); 32348 (15.6.4); 32365 (5.3.1); 32379 (26.10.4); 32381 (6.1.16); 32383 (16.7.3); 32383A (26.7.3); 32386 (13.2.1); 32387 (15.4.4); 32388 (23.16.1); 32389 (20.2.3); 32393 (26.13.15); 32395 (16.8.6); 32398 (26.13.10a); 32403 (24.1.13); 32407 (15.4.5, 15.4.22); 32410 (8.1.3); 32412 (24.1.9); 32415 (9.5.1); 32416 (15.4.5); 32417 (7.1.8); 32444 (24.1.6); 32446 (15.3.5); 32447 (23.8.3); 32454 (28.6.17); 32460 (6.1.34); 32461 (10.10.3a); 32463 (15.4.11, 15.4.20); 32464 (26.13.2); 32465 (26.10.4); 32466 (12.1.1); 32467 (15.2.9, 23.18.1); 32467A (15.3.5); 32473 (15.1.1); 32484 (5.6.1); 32485 (26.11.7b); 32485a (26.11.4); 32486 (16.2.1); 32512 (7.1.1); 32513 (16.1.1, 16.14.1); 32514 (24.1.2); 32516 (28.6.17); 32517 (13.6.1); 32518 (28.2.2); 32521 (5.1.1); 32527 (21.3.2b); 32546 (26.6.1); 32547 (8.1.15); 32548A (26.13.14); 32551 (28.6.17); 32552 (20.3.3); 32562 (24.1.14); 32563 (28.6.2); 32564 (26.2.1); 32565 (6.1.5); 32569 (20.2.3); 32570 (15.4.4); 32571 (28.4.1); 32572 (8.1.4); 32591 (6.1.28); 32593 (26.13.15); 32593a (26.10.1); 32594 (26.10.1); 32607 (28.6.4); 32636 (22.1.2); 32644 (9.4.4); 32653 (13.9.4); 32716 (16.8.3); 32738 (24.1.13); 32751 (16.8.2); 32759 (15.4.5, 15.4.16, 15.4.19); 32759A (15.4.10); 32810 (10.4.14); 32811 (15.6.2b); 32830 (17.5.1); 32869 (15.3.5); 32871 (20.3.3); 32925 (2.3.1, 2.3.5); 32930 (5.3.1); 32951 (16.12.2); 32952 (28.6.17); 32957 (2.1.5); 32958 (15.2.1); 33029 (6.1.22); 33055 (10.4.17c); 33090 (26.13.5); 33101 (15.4.5); 33103 (15.4.5); 33122 (7.1.4); 33123 (15.3.21); 33124 (15.4.6); 33140 (6.1.16); 33154 (8.1.10); 33156 (8.1.18); 33158 (26.13.7); 33160 (15.6.4); 33161 (15.2.2); 33162 (16.8.6); 33163 (28.6.3); 33181 (9.4.4); 33205 (15.3.8); 33206 (15.5.3); 33208 (10.7.1); 33209 (5.4.1); 33210 (15.4.6); 33211 (5.4.1); 33212 (15.3.4); 33213 (16.8.1); 33218 (16.7.2); 33222 (13.13.2); 33224 (15.3.20); 33229 (23.20.5); 33230 (23.2.1); 33283 (8.1.13); 33666 (15.4.13); 33675 (26.4.2a); 33677 (15.1.1); 33680 (6.1.16); 33681 (23.4.1); 33682 (28.6.17); 33683 (26.4.2a); 33684 (13.6.1); 33692 (23.18.1); 33695 (26.6.2); 33696 (10.10.3a); 33702 (26.11.7a); 33716 (15.5.2); 33718 (16.7.6); 33719 (26.9.1); 33723 (28.6.13); 33724 (10.4.14); 33725 (7.1.7); 33737 (23.2.3a); 33738 (15.3.21); 33739 (15.4.6); 33740 (10.4.14); 33742 (15.4.6); 33743 (15.4.22); 33744 (6.1.10); 33745 (28.2.1); 33746 (23.8.4); 33747 (28.6.2); 33748 (28.1.1); 33750 (13.11.1); 33753 (13.9.8); 33789 (15.4.13); 33810 (28.6.7a); 33821 (13.3.3); 33827 (6.1.31); 33865 (18.1.4); 33892 (16.8.2); 33893 (16.7.3); 33927 (16.10.4); 33927A (15.4.13); 33955 (13.9.5); 33965 (15.4.5); 33980 (23.19.2); 33984 (16.1.1); 33999 (26.13.2); 34012 (7.1.8, 8.1.1); 34013 (24.1.9); 34030 (2.2.1); 34041 (9.3.2); 34044 (28.4.1); 34104 (16.7.6); 34105 (26.2.1); 34108 (10.4.17b); 34133 (26.12.4); 34134 (13.6.1); 34137 (11.1.1); 34138 (28.4.1); 34140 (10.9.1a); 34142 (24.1.14); 34195 (16.12.2); 34196 (16.8.12); 34197 (9.4.3); 34222 (15.1.1); 34228 (8.1.18); 34279 (15.7.1); 34348 (15.1.1); 34372 (16.12.2); 34392 (23.22.1); 34394 (15.4.5); 34396 (28.6.18); 34397 (10.3.4); 34468 (18.1.4); 34476 (28.6.28); 34477 (16.8.4); 39161 (15.2.7); 40013 (15.1.1); 40014 (16.14.1); 40016 (15.4.12); 40040 (16.14.1); 40051 (16.7.3); 40142 (15.3.19, 8.1.3); 40156 (15.5.2); 40176 (23.10.1); 40189 (14.3.2); 40292 (10.4.14); 40294 (27.1.3); 40295 (17.2.6); 40296 (23.13.11); 40306 (26.12.1); 40307 (24.1.11); 40308 (8.1.20); 40309 (26.10.1); 40311 (10.1.3); 40314 (2.1.6); 40354 (8.1.20); 40564 (26.13.5); 40565 (28.6.8); 40594 (26.11.2); 40650 (13.7.1); 40652 (13.16.9); 40694 (23.16.4); 40792 (15.4.20); 40806 (28.6.20); 40812 (7.1.2); 40837 (15.7.1); 40877 (16.1.1); 40900 (16.7.6); 40962p.p. (15.4.12); 40963 (27.1.3); 40983 (18.1.5); 50029 (13.6.1); 50133 (15.4.4); 50263 (17.3.1); 50319 (2.1.3); 50396 (8.1.20); 50410 (18.2.1); 50453 (23.9.1); 50482 (6.1.19a); 50578 (26.13.5); 50583 (15.6.4); 50592 (17.5.1); 50593 (10.4.18); 50681 (16.8.5); 50697 (16.8.6); 50708 (10.3.2); 50709 (11.1.1); 50733 (6.1.34); 50753 (7.1.12); 50789 (2.3.1); 50803 (2.1.1); 50809 (13.13.2); 50857 (6.1.13); 50858 (13.1.1); 50860 (28.6.13); 50864 (13.13.3); 50865 (15.2.10); 50890 (16.10.4); 50893 (16.7.3); 50894 (16.7.2); 50896 (26.4.2a); 50897 (6.1.10); 50900 (15.5.3); 50901 (15.3.12); 50906 (15.6.2b); 50907 (28.2.1); 50908 (10.7.1); 50909 (15.4.6); 50911 (15.4.22, 9.4.4); 50919 (23.4.5); 50921 (15.3.21); 50928 (2.3.6); 50930 (10.7.1); 50959 (15.3.20); 51040 (23.4.1); 51110 (15.4.13); 51111 (15.4.13); 51112 (22.1.2); 51113 (16.8.5); 51115 (16.10.1); 51125 (15.3.12); 51126 (15.3.20); 51187 (7.1.5); 51188 (8.1.13); 51190 (15.4.14); 51191 (15.2.2); 51245 (8.1.12); 51308 (17.3.1); 51357 (2.1.5); 51364 (23.2.3a); 51366 (9.4.4); 51367 (15.4.21); 51368 (15.6.2b); 51368a (15.6.2b); 51391 (13.13.3); 51392 (13.15.1); 51394 (2.1.1); 51396 (23.2.3a); 51397 (16.8.6); 51449 (17.2.1); 51450 (7.1.4); 51504 (13.13.1); 51510 (2.3.3); 51539 (13.8.1); 51542 (13.13.3); 51543 (16.7.4); 51629 (2.3.2); 51630 (13.13.5); 51699 (9.3.2); 51701 (23.8.2); 51702 (15.3.1); 51704 (9.3.2); 51709 (15.2.3); 51710 (24.1.12); 51757 (8.1.10); 51760 (7.1.10); 51761 (25.2.4a).

Collenette 540 (16.12.2); 544 (2.3.4); 561 (2.3.3); 562 (14.2.2); 563 (9.4.4); 614 (17.2.1); 615 (15.4.17); 617 (15.3.8); 618 (15.5.3); 899 (16.12.2); 900 (16.15.1); 1022 (15.2.10); 1029 (23.19.2); 1034 (5.1.1); 21505 (10.7.1); 21507 (2.3.2); 21509 (23.16.2); 21510 (2.3.4); 21513 (13.15.1); 21516 (7.1.4); 21517 (22.1.2); 21521 (10.2.1); 21527 (17.2.7); 21528 (23.4.5); 21542 (16.5.1); 21544 (15.3.21); 21546 (6.1.10); 21548 (5.4.1); 21554 (12.2.2); 21555 (23.16.2); 21556 (26.4.2a); 21557 (14.2.2); 21558 (14.4.1a); 21561 (2.3.5); 21611 (24.1.13); 21612 (26.13.7); 21620 (23.16.4); 21628 (21.3.3a); 21651 (24.1.16); 21655 (6.1.16); 21661 (2.1.7).

Cox 2527 (15.3.4); 2530 (16.8.2); 2549 (9.4.4); 2550a (15.2.3).

Edwards 2160 (10.4.14); 2163 (4.1.3); 2164 (16.14.1); 2165 (10.4.13); 2166 (15.4.20); 2168A (16.8.7); 2168B (16.7.7); 2173 (26.13.1); 2174 (28.6.11); 2175 (15.5.2); 2176C (15.7.1); 2176D (15.7.2); 2179 (28.4.1); 2180 (4.1.5); 2182 (10.4.13); 2183 (10.4.13); 2184 (23.19.2); 2185 (4.1.3); 2187 (8.1.9); 2188

(8.1.7); 2189 (22.1.2); 2190 (15.4.6); 2191 (28.2.1); 2192 (15.6.2); 2193A (15.4.8); 2193B (15.4.6); 2194 (13.13.2); 2195 (15.7.3); 2196 (8.1.11); 2198 (8.1.9); 2199 (8.1.14); 2200 (14.2.2); 2201 (16.7.3); 2202 (6.1.10); 2203 (23.4.5); 2204 (15.4.13); 2205 (25.2.2); 2208 (26.10.1); 2210 (13.13.2); 2211 (16.8.1); 2212 (15.3.14); 2213 (15.3.8); 2215 (15.4.4); 2216 (15.3.4); 2218 (15.4.29) & (15.7.1); 2219 (15.4.13); 2220 (15.3.4); 2222 (16.10.5); 2229 (15.7.1); 2237 (14.4.1a).

Enriquez 18163 (15.1.1).

Fuchs 21463 (16.8.5); 21464 (16.7.3); 21465 (15.3.21); 21466 (15.4.7); 21468 (6.1.10); 21470 (5.4.1); 21474 (15.3.20); 21475 (26.12.2); 21477 (26.9.1); 21479 (8.1.9); 21480 (13.9.8); 21481 (13.13.5); 21484 (26.12.2); 21485 (23.16.2); 21488 (22.1.3); 21490 (14.4.1a).

Fuchs & Collenette 21444 (2.3.4); 21668 (28.4.1).

Gibbs 401 (16.4.3); 3953 (20.2.3); 3962 (16.8.5); 3966 (14.4.4a); 3992 (12.1.1); 3998 (12.1.1); 3999 (11.1.1); 4017 (15.1.1); 4018 (15.2.2, 15.7.1); 4020 (16.8.4); 4036 (16.16.2); 4062 (15.4.19); 4068 (15.6.4); 4092 (5.4.1); 4140 (15.2.1); 4141 (15.4.7); 4143 (16.12.1); 4144 (16.8.5); 4228 (16.8.3); 4237 (8.1.11); 4251 (15.3.4); 4253 (16.10.3); 4255 (15.4.14); 4265 (15.4.6); 4272 (15.4.17); 4299 (15.7.3).

Hale 28266 (16.15.1); 29040 (15.2.2).

Haviland 1353 (1.1.1a); 1410 (2.1.1); 1411 (2.1.1); 1412 (2.3.6); 1413 (2.3.4); 1414 (2.3.1); 1415 (4.1.11); 1416 (2.1.5); 1474 (23.19.2); 1475 (28.4.1); 1476 (18.1.4); 1477 (6.1.5); 1478 (16.8.11); 1479 (8.1.14); 1480 (6.1.10); 1481 (15.6.2b); 1482 (15.2.10); 1483 (14.3.4); 1484 (15.5.3); 1485 (8.1.7); 1486 (26.4.2a); 1488 (15.4.8); 1489 (23.4.4); 1490 (23.16.2); 1491 (9.4.3); 1492 (10.2.1); 1494 (6.1.9); 1495 (10.4.9).

Holttum 1 (26.13.1); 4 (26.13.15); 5 (26.4.3); 13 (26.4.2a); 19 (26.13.2); 20 (26.6.2); 22 (26.13.5); 23 (26.11.6); 24 (26.10.4); 28 (17.2.6); 29 (17.2.4); 34 (26.13.3); 37 (26.13.9); 39 (23.3.1); 44 (26.2.2a); 45 (17.2.8); 48 (12.1.1); 52 (26.2.2a); 53 (26.7.3); 56 (26.13.14); 57 (26.12.2); 58 (26.10.3); 59 (17.2.4); 25106 (11.3.1); 25107 (13.16.2); 25108 (13.16.7); 25121 (26.3.4); 25122 (6.1.35); 25123 (23.20.3); 25130 (26.13.12); 25131 (4.1.4); 25132 (10.4.5); 25133 (4.1.3); 25134 (28.6.29a); 25138 (10.5.1); 25139 (28.6.7a); 25140 (28.6.7a); 25141 (17.1.2); 25142 (16.3.3); 25143 (23.22.1); 25144 (26.11.3); 25145 (10.4.17a); 25146 (6.1.34); 25147 (26.11.5); 25148 (23.13.9); 25149 (10.4.2); 25150 (27.1.7); 25179 (15.6.4); 25251 (17.3.1); 25252 (4.1.3); 25253 (13.4.3a); 25254 (26.6.1); 25257 (13.10.1); 25258 (28.6.30); 25259 (28.6.16); 25260 (26.10.4); 25261 (13.14.1); 25262 (4.1.8); 25263 (26.13.16a); 25269 (21.3.3a); 25274 (23.13.4); 25275 (24.1.5); 25276 (1.1.1a); 25277 (5.3.1); 25289 (10.1.2); 25290 (5.1.1); 25291 (23.11.2); 25292 (23.2.1); 25293 (23.16.1); 25294 (26.5.1); 25295 (9.4.3); 25296 (10.5.3); 25297 (23.13.11); 25298 (10.1.3); 25299 (10.3.4); 25300 (10.3.4); 25351 (16.8.4); 25352 (17.1.2); 25353 (9.1.1); 25355 (6.1.22); 25356 (23.11.3); 25356b (23.11.1); 25357 (17.2.6); 25358 (23.20.5); 25359 (13.3.2); 25360 (6.1.36); 25361 (4.1.11); 25362 (27.1.1); 25363 (24.1.7); 25364 (6.1.1); 25365 (6.1.5); 25366 (15.5.1); 25367 (4.1.4); 25368 (4.1.5); 25369 (26.5.2); 25375 (16.13.2); 25377 (10.8.1); 25378 (11.1.1); 25379 (28.6.24); 25380 (28.6.23); 25381 (10.5.2); 25382 (18.1.4); 25383 (6.1.28); 25384 (26.7.3); 25385 (26.13.1); 25386 (28.6.4); 25387 (13.9.3); 25388 (26.11.7c); 25389 (26.13.2); 25390 (13.9.5); 25391 (18.3.1); 25392 (16.7.8); 25393 (23.8.3); 25394 (13.13.1); 25395 (26.13.9); 25396 (16.1.1); 25397 (4.1.3); 25398 (4.1.12); 25399 (15.4.20); 25400 (16.3.2); 25401 (16.14.1); 25402 (6.1.16); 25403 (10.4.13); 25404 (13.16.4); 25405 (14.2.4); 25406 (10.1.1); 25407 (8.1.8); 25408 (8.1.16); 25409 (8.1.21); 25410 (26.11.2); 25411 (10.3.3); 25412 (17.2.4); 25413 (24.1.12); 25414 (10.2.1); 25415 (9.2.2); 25417 (26.4.3); 25418 (23.16.4); 25419 (27.3.2); 25420 (10.4.4); 25423 (6.1.21); 25424 (16.7.2); 25425 (16.8.5); 25426 (16.8.12); 25427 (17.2.8); 25428 (15.2.3); 25429 (28.6.3); 25430 (16.14.1); 25431 (15.3.18); 25432 (15.3.6); 25433 (10.4.4); 25434 (16.4.3); 25435 (10.4.12); 25436 (4.1.3); 25437 (27.3.1); 25438 (14.1.3a); 25439 (13.3.3); 25440 (26.6.1); 25441 (23.4.2); 25442 (22.1.1); 25443 (7.1.12); 25444 (28.6.26); 25445 (11.4.1); 25447 (8.1.3); 25448 (14.2.3); 25449 (14.4.1a); 25450 (14.2.2); 25451 (12.1.1); 25452 (23.4.1); 25453 (23.4.5); 25454 (17.2.2); 25455 (2.3.2); 25457 (13.9.4); 25458 (8.1.14); 25459 (10.4.18); 25460 (16.15.1); 25461 (15.3.4); 25462 (16.10.4); 25464 (15.6.4); 25465 (15.2.1); 25467 (15.7.2); 25468 (16.12.1, 16.12.2); 25469 (16.6.1); 25470 (16.8.11); 25471 (26.4.4); 25472 (26.4.2a); 25473 (15.1.1); 25474 (10.7.1); 25475 (16.10.1); 25476 (15.7.3); 25477 (15.4.6); 25478 (15.4.8); 25479 (15.4.13); 25480 (15.4.8); 25482 (16.8.8); 25483 (15.4.13); 25484 (15.4.3); 25485 (6.1.29); 25486 (28.2.3); 25487 (23.2.3a); 25488 (13.8.1); 25489 (2.1.1); 25490 (13.13.3); 25491 (22.1.4); 25496 (15.4.15); 25497 (16.7.4); 25498 (16.4.5); 25499 (13.1.1); 25500 (15.2.2); 25501 (16.7.3); 25502 (6.1.10); 25503 (15.4.22); 25504 (23.11.4); 25505 (25.2.3); 25506 (15.5.3); 25507 (28.2.4); 25508 (17.2.1); 25509 (8.1.7); 25510 (14.3.4); 25511 (2.3.4); 25512 (26.4.2a); 25513 (13.13.2); 25514 (13.8.1); 25515 (23.16.2); 25517 (13.4.3a); 25518 (28.6.20); 25519 (10.2.1); 25520 (10.1.4a); 25521 (24.1.6); 25522 (28.6.29); 25523 (26.4.3); 25524 (15.3.5); 25524b (15.3.21); 25525 (13.3.3); 25525b (13.13.4); 25526 (28.2.2); 25527 (8.1.19); 25528 (7.1.8); 25529 (28.6.13); 25530 (28.2.1); 25532 (15.4.7); 25533 (15.4.5); 25534 (13.13.4); 25535 (10.3.2); 25536 (27.1.7); 25537 (6.1.31); 25538 (26.14.1); 25541 (16.16.2); 25542 (15.6.2b); 25543 (23.21.2); 25544 (5.4.1); 25545 (13.12.3); 25546 (10.10.3a); 25547 (23.8.4); 25548 (6.1.15); 25549 (9.4.3); 25550 (15.3.18); 25551

(6.1.9); 25552 (18.1.5); 25553 (5.1.1); 25554 (16.3.1); 25555 (28.6.16); 25556 (24.1.3); 25557 (13.9.5); 25558 (24.1.11); 25559 (28.6.21); 25560 (8.1.4); 25561 (13.13.1); 25562 (28.6.1); 25563 (2.1.6); 25564 (10.5.5); 25565 (28.6.27); 25566 (13.3.1); 25567 (9.3.2); 25568 (23.13.5); 25569 (24.1.9); 25570 (13.16.8); 25571 (13.16.6); 25572 (23.13.1); 25573 (9.3.1); 25574 (26.3.3); 25575 (26.3.4); 25576 (13.9.6); 25577 (28.4.1); 25579 (28.6.25); 25580 (16.8.12); 25581 (28.5.1); 25582 (6.1.1); 25583 (28.4.2); 25584 (13.14.1); 25585 (17.1.1); 25586 (6.1.31); 25588 (5.3.2); 25589 (23.13.11); 25592 (26.11.6); 25593 (23.7.1); 25594 (23.8.4); 25595 (24.1.1); 25596 (26.11.3); 25597 (10.4.15); 25598 (13.16.7); 25599 (26.6.1); 25600 (26.13.15); 25601 (8.1.10); 25602 (10.4.8); 25603 (10.10.4); 25604 (15.3.16); 25605 (16.8.3); 25606 (26.13.7); 25607 (10.4.9); 25608 (10.4.12); 25609 (26.4.2a); 25611 (2.1.7); 25613 (15.4.13, 15.6.1); 25614 (24.1.8); 25615 (6.1.20); 25616 (28.6.14); 25617 (14.1.3b); 25618 (16.8.12); 25619 (19.1.1); 25620 (4.1.6); 25621 (24.1.13); 25624 (23.12.1); 25625 (8.1.8); 25626 (13.5.1); 25627 (16.3.2); 25628 (23.13.3); 25630 (28.6.19); 25714 (13.13.4); 25715 (27.2.1); 25716 (18.1.1); 25717 (17.2.8); 25718 (17.2.7); 25720 (23.13.3); 25722 (24.1.2); 25723 (20.3.2); 25724 (9.4.4); 25725 (9.4.2); 25726 (13.16.10); 25727 (26.3.2); 25728 (10.4.4); 25729 (10.4.13); 25730 (22.1.2); 25731 (15.3.7); 25732 (15.2.10); 25733 (16.12.2); 25734 (16.7.3); 25735 (16.8.11); 25736 (15.6.4); 25737 (15.4.1, 15.4.20); 25738 (15.4.20); 25739 (15.4.11); 25740 (15.7.4); 25741 (15.4.14); 25742 (15.4.4); 25743 (15.4.20); 25744 (15.4.13); 25745 (15.4.13); 25870 (15.4.6); 25871 (15.4.21); 25872 (15.4.29); 25873 (15.4.13).

Hotta 3789 (16.7.3); 3825 (16.7.2); 3850 (16.10.2); 20318 (26.10.1); 20449 (8.1.4).

Jacobs 5021 (8.1.17); 5701 (8.1.17); 5712 (15.4.29); 5713 (15.4.8); 5714 (15.2.7); 5716 (15.4.6); 5717 (8.1.11); 5718 (15.7.3); 5721 (15.6.2b); 5725 (14.3.4); 5728 (14.2.2); 5729 (15.3.4); 5732 (25.2.3); 5735 (15.4.13); 5738 (15.4.17, 15.4.22); 5739 (15.4.8); 5740 (15.4.13); 5742 (15.5.3); 5758 (15.4.5); 5767 (15.3.3); 5770 (26.4.2a); 5773 (8.1.11); 5774 (15.2.8); 5778 (8.1.13); 5780 (28.6.13); 5787 (8.1.15); 5788 (8.1.1); 5789 (8.1.14); 5790 (8.1.3); 5792 (15.2.8); 5793 (15.4.22); 5794 (15.4.13); 5795 (15.3.4); 5796 (15.4.14); 5797 (15.4.8).

Kanis 51463 (10.9.2); 51471 (15.4.19); 53975 (16.8.6); 53988 (16.14.1).

Kitayama 64 3817 (15.4.8); Kitayama K665 3828 (15.3.6); Kitayama K716 3877 (15.4.22); Kitayama K742 3903 (15.4.13); Kitayama K743 3904 (15.3.8).

Kodama 12691 (23.4.2).

Kokawa 6312 (16.8.12).

Kokawa & Hotta 2836 (16.15.1); 2894 (16.1.1); 2942 (16.8.6); 2944 (16.14.1); 3035 (26.5.2); 3228 (16.15.1); 3229 (16.8.5); 3233 (16.8.1); 3455 (26.4.2a); 3526 (16.4.3); 3699 (16.16.2); 3731 (16.8.13); 3972 (16.7.6); 4029 (16.16.2); 4076 (16.7.2); 4273 (16.7.3); 4329 (26.2.1); 4434 (16.7.6); 4492 (26.10.4); 4535 (26.4.3); 4734 (26.10.4); 4836 (16.3.2); 4837 (16.4.3); 4869 (23.11.1); 4922 (26.13.11); 4930 (26.8.1a); 5151 (16.15.1); 5267 (16.12.2); 5268 (16.15.1); 5435 (26.4.2a); 5505 (16.15.1); 5556 (16.14.1); 5557 (16.12.2); 5822 (16.7.6).

Meijer 20267 (13.4.4); 20314 (18.1.4); 20989 (10.10.1); 22032 (15.4.13); 22048 (2.3.4); 22061 (8.1.7); 22062 (15.5.3); 22066 (28.2.1); 22071 (14.3.4); 22072 (14.2.2); 22078 (23.16.2); 22079 (15.4.6); 23842 (2.3.1); 24039 (16.14.1); 24042 (8.1.12); 24059 (11.1.1); 24071 (10.10.3a); 24122 (14.2.2, 14.4.1a); 24134 (13.1.1); 24138 (7.1.5, 26.4.2a); 28578 (8.1.7); 29113 (8.1.9); 29117 (2.1.1); 29122 (13.13.2); 29150 (8.1.14); 29188 (11.4.1); 29214 (16.16.2); 29215 (16.12.2); 29252 (2.3.4); 29272 (15.5.3); 38586 (2.3.5); 38590 (14.2.1); 38593 (9.4.4); 38594 (8.1.11); 38595 (14.4.1a).

Mikil 29225 (14.2.2).

Molesworth-Allen 3233 (26.8.1a); 3235 (26.7.2); 3241 (16.15.1); 3245 (16.8.13); 3250 (23.2.3a); 3256 (15.3.21); 3257 (15.3.21); 3258 (15.3.4); 3259 (15.6.2b); 3264 (2.3.2); 3265 (15.2.10); 3268 (13.15.1); 3270 (22.1.3); 3273 (6.1.2); 3282 (15.3.8); 3283 (7.1.4); 3291 (2.1.1); 3303 (6.1.10).

Parris 85/45 (4.1.14); 85/46 (4.1.13); 85/47 (2.2.1); 10780 (23.20.2); 10781 (15.7.1); 10782 (16.4.3); 10783 (16.7.7); 10784 (16.8.11); 10785 (16.8.7); 10786 (10.4.12); 10787 (8.1.3); 10788 (10.5.2); 10789 (8.1.20); 10790 (26.13.5); 10791 (23.16.4); 10792 (26.7.3); 10793 (24.1.12); 10794 (28.6.7b); 10795 (26.7.1); 10796 (13.12.3); 10797 (26.13.5); 10798 (26.7.3); 10799 (17.2.2); 10800 (20.3.1); 10801 (26.4.4); 10802 (9.2.2); 10803 (26.4.3); 10804 (26.4.2a); 10805 (26.7.1); 10806 (10.1.5); 10807 (8.1.8); 10808 (8.1.15); 10809 (28.6.17); 10810 (15.3.18); 10811 (23.18.1); 10812 (16.7.2); 10813 (16.14.1); 10814 (16.4.2); 10815 (16.8.6); 10816 (16.7.7); 10817 (16.8.1); 10818 (16.8.2); 10819 (16.8.3); 10820 (16.8.4); 10821 (16.8.11); 10822 (13.9.6); 10823 (6.1.16); 10824 (20.3.1); 10825 (10.4.17c); 10826 (10.1.1); 10827 (8.1.8); 10828 (26.13.15); 10829 (16.7.9); 10830 (16.8.12); 10831 (26.7.2); 10832 (26.3.2); 10833 (17.2.6); 10834 (28.6.19); 10835 (26.13.15); 10836 (28.6.7b); 10837 (26.13.1); 10838 (6.1.8); 10839 (26.10.1); 10840 (26.4.2a); 10841 (16.8.7); 10842 (26.6.1); 11421 (16.4.3); 11422 (28.6.28); 11423 (16.7.7); 11424 (4.1.3); 11425 (26.7.1); 11426 (16.4.2); 11427

(27.1.8); 11428 (4.1.14); 11429 (2.1.5); 11430 (4.1.13); 11431 (28.6.26); 11432 (15.2.6); 11433 (15.4.20); 11434 (15.4.19); 11435 (16.14.1); 11436 (16.8.11); 11437 (13.16.4); 11438 (10.4.8, 10.5.1); 11439 (26.10.1); 11440 (26.7.2); 11442 (13.3.3); 11443 (10.1.4a); 11444 (16.8.12); 11445 (26.4.2a); 11446 (16.8.1); 11447 (16.8.1); 11448 (28.1.1); 11449 (24.1.12); 11450 (16.7.2); 11451 (28.2.5); 11452 (26.7.2); 11453 (6.1.10); 11454 (26.13.15); 11455 (7.1.1); 11456 (16.7.2); 11457 (13.13.4); 11458 (26.8.1a); 11459 (26.12.1); 11460 (4.1.5); 11461 (26.7.3); 11462 (10.4.14); 11463 (23.3.1); 11464 (16.3.2); 11465 (16.8.11); 11466 (22.1.2); 11467 (8.1.14); 11468 (16.10.3); 11469 (10.4.4); 11470 (7.1.5); 11471 (16.14.1); 11472 (28.6.24); 11473 (27.3.1); 11474 (16.12.2); 11475 (5.4.1); 11476 (6.1.21); 11477 (10.5.2); 11478 (23.10.1); 11479 (28.3.1); 11480 (23.11.3); 11481 (6.1.16); 11482 (6.1.16); 11483 (28.6.29a); 11484 (28.6.2); 11485 (4.1.12); 11486 (26.13.5); 11487 (6.1.7); 11488 (16.1.1); 11489 (6.1.16); 11490 (28.1.1); 11491 (13.9.3); 11492 (15.3.15); 11493 (15.4.4); 11494 (28.4.1); 11495 (6.1.34); 11496 (16.8.3); 11497 (16.4.3); 11498 (9.4.3); 11499 (15.7.1); 11500 (11.2.1); 11501 (16.8.1); 11502 (16.8.6); 11503 (23.16.3a); 11504 (15.3.6); 11505 (10.10.3a); 11506 (28.6.23); 11507 (28.6.11); 11508 (26.7.1); 11509 (10.9.1a); 11510 (2.2.1); 11511 (16.10.3); 11512 (16.4.4); 11513 (12.1.1); 11514 (15.2.3); 11515 (15.3.18); 11516 (13.9.6); 11517 (11.2.1); 11518 (26.11.7b); 11519 (24.1.13); 11520 (26.13.4); 11521 (4.1.14); 11522 (8.1.10); 11523 (10.10.4); 11524 (16.14.1); 11525 (15.2.3); 11526 (15.4.25); 11527 (10.4.9); 11528 (10.10.2); 11529 (4.1.14); 11530 (4.1.2); 11531 (6.1.5); 11532 (6.1.28); 11533 (18.3.1); 11534 (15.6.4); 11535 (13.6.1); 11536 (13.3.3); 11537 (16.10.1); 11538 (16.12.2); 11539 (26.4.2a); 11540 (13.1.1); 11541 (22.1.2); 11542 (28.2.2); 11543 (14.3.1); 11544 (14.4.3a); 11545 (10.7.1); 11546 (28.6.26); 11547 (10.3.2); 11548 (13.9.4); 11549 (23.16.2); 11550 (23.4.1); 11551 (9.4.4); 11552 (15.2.3); 11553 (14.2.2); 11554 (23.4.5); 11555 (15.3.12); 11556 (16.8.6); 11557 (15.4.9); 11558 (15.6.4); 11559 (26.13.8); 11560 (10.1.5); 11561 (2.1.5); 11562 (8.1.7); 11563 (13.13.2); 11564 (15.3.4); 11565 (22.1.2); 11566 (7.1.9); 11567 (15.6.2a); 11568 (16.7.4); 11569 (25.2.3); 11570 (28.2.1); 11571 (6.1.10); 11572 (13.8.1); 11573 (13.9.8); 11574 (13.8.1); 11575 (23.2.3a); 11576 (23.11.4); 11577 (13.13.2); 11578 (17.2.1); 11579 (7.1.4); 11580 (16.8.10); 11581 (15.5.3); 11582 (10.2.1); 11583 (13.8.1); 11584 (15.3.8); 11585 (15.4.17); 11586 (16.8.10); 11587 (6.1.2); 11588 (13.13.2); 11589 (15.4.6); 11590 (15.2.10); 11591 (15.4.15); 11592 (6.1.13); 11593 (28.2.3); 11594 (15.3.14); 11595 (13.15.1); 11596 (6.1.29); 11597 (15.3.21); 11598 (28.6.26); 11599 (8.1.13); 11600 (16.10.1); 11601 (12.2.2); 11602 (16.10.5); 11603 (2.3.2); 11604 (2.3.6); 11605 (14.2.1); 11606 (16.12.1); 11607 (16.7.3); 11608 (28.2.4); 11609 (24.1.6); 11610 (26.13.14); 11611 (16.8.9); 11612 (10.4.18); 11613 (16.8.6); 11614 (9.1.1); 11615 (14.1.3a); 11616 (14.1.3d); 11617 (26.7.1); 11618 (16.15.1); 11619 (26.4.4); 11620 (14.1.3b); 11621 (10.8.1); 11622 (26.11.2); 11623 (17.2.8); 11646 (21.3.3a); 11659 (23.4.2); 11660 (15.6.2b); 11661 (15.2.8); 11662 (15.4.19); 11663 (15.2.6); 11664 (15.4.8); 11665 (15.4.13); 11666 (15.4.17); 11667 (15.4.29); 11668 (2.1.1); 11669 (2.1.1); 11670 (15.3.8); 11671 (16.4.6); 11672 (16.8.7); 11673 (15.3.11).

Parris & Croxall 8468 (15.4.8); 8469 (15.7.1); 8470 (15.7.2); 8471 (15.3.18); 8472 (15.3.6); 8473 (15.5.2); 8474 (23.18.1); 8475 (15.6.2b); 8476 (15.3.4); 8477 (15.4.20); 8478 (15.2.3); 8479 (15.3.15); 8480 (15.4.4); 8481 (15.3.11); 8482 (15.4.24); 8483 (15.4.12); 8484 (15.2.4); 8485 (15.3.19); 8486 (15.4.7); 8487 (15.3.21); 8488 (15.2.2); 8489 (15.6.4); 8490 (15.4.6); 8491 (15.4.14); 8492 (15.4.11); 8493 (15.3.20); 8494 (15.2.4); 8495 (15.3.12); 8496 (15.2.1); 8497 (15.1.1); 8498 (15.1.2); 8499 (15.3.4); 8500 (15.4.29); 8501 (15.4.25); 8502 (15.4.9); 8503 (15.4.17); 8504 (15.3.8); 8505 (15.5.3); 8506 (15.2.10); 8507 (15.3.14); 8508 (23.11.4); 8509 (15.4.21); 8510 (15.4.6); 8511 (15.4.22); 8512 (15.4.8); 8513 (15.4.13); 8514 (15.4.15); 8515 (15.4.3); 8516 (15.4.3); 8517 (15.2.10); 8518 (15.2.6); 8519 (15.6.2a); 8520 (23.11.2); 8521 (15.3.3); 8522 (15.4.18); 8523 (15.3.7); 8524 (15.4.5); 8525 (15.4.26); 8526 (15.4.16); 8527 (23.11.3); 8528 (15.3.5); 8529 (15.4.2); 8530 (15.4.27); 8531 (15.4.28); 8532 (26.7.3); 8533 (26.7.1); 8534 (26.13.1); 8535 (26.7.1); 8536 (26.13.15); 8537 (26.4.3); 8538 (26.13.5); 8539a (26.11.7c); 8539b (26.11.7b); 8540 (9.1.1); 8541 (6.1.21); 8542 (6.1.16); 8543 (6.1.16); 8544 (22.1.1); 8545 (28.6.23); 8546 (28.6.28); 8547 (28.6.7b); 8548 (11.2.1); 8549 (11.4.1); 8550 (10.2.2); 8551 (20.3.1); 8552 (14.1.3b); 8553 (10.4.8); 8554 (10.4.17c); 8555 (10.4.12); 8556 (10.4.12); 8557 (16.8.12); 8558 (16.8.6); 8559 (16.8.11); 8560 (26.10.1); 8561 (26.7.2); 8562 (28.6.26); 8563 (23.2.2); 8564 (23.4.2); 8565 (15.6.2b); 8566 (15.2.9); 8567 (23.2.1); 8568 (26.10.4); 8569 (10.5.1); 8570 (13.9.6); 8571 (28.6.24); 8572 (23.4.2); 8573 (23.19.2); 8574 (16.15.1); 8575 (16.6.1); 8576 (16.14.1); 8577 (16.6.2); 8578 (16.10.3); 8579 (16.7.2); 8580 (16.12.2); 8581 (16.8.2); 8582 (16.8.2); 8583 (16.8.7); 8584 (16.8.7); 8585 (16.8.11); 8586 (16.8.4); 8587 (16.8.4); 8588 (16.8.2); 8589 (27.2.1); 8590 (27.3.1); 8591 (9.4.3); 8592 (9.4.3); 8593 (10.4.10); 8594 (10.4.14); 8595 (28.4.1); 8596 (6.1.32); 8597 (6.1.31); 8598 (6.1.34); 8599 (6.1.34); 8600 (6.1.16); 8601 (6.1.16); 8602 (23.8.3); 8603 (26.11.3); 8604 (13.2.1); 8605 (23.3.3); 8606 (23.13.11); 8607 (27.1.7); 8608 (17.2.3); 8609 (13.9.3); 8610 (12.1.1); 8611 (23.10.1); 8612 (13.2.1); 8614 (16.14.1); 8615 (16.7.2); 8616 (16.4.6); 8617 (16.4.6); 8618 (16.13.1); 8619 (16.12.2); 8620 (16.3.2); 8621 (16.4.3); 8622 (16.7.7); 8623 (16.7.9); 8624 (16.7.8); 8625 (16.8.11); 8626 (16.7.7); 8627 (16.8.4); 8628 (16.8.7); 8629 (16.8.7); 8630 (16.8.5); 8631 (16.8.6); 8632 (16.8.2); 8633 (16.8.2); 8634 (15.4.20); 8635 (15.2.3); 8636 (15.2.11); 8637 (15.2.6); 8638 (15.2.11); 8639 (15.2.6); 8640 (15.2.6); 8641 (15.2.2); 8642 (15.4.4); 8643 (15.4.22); 8644 (15.3.14); 8645 (15.5.2); 8646 (15.2.12); 8647 (15.4.29); 8648 (15.6.4); 8649 (15.3.12); 8650 (15.2.4); 8651 (15.2.3); 8652 (15.4.4); 8653 (15.7.2); 8654 (15.6.2b); 8655 (15.6.2b); 8656 (15.2.6); 8657 (15.4.29); 8658 (15.4.12); 8659 (15.4.25); 8660 (15.4.9); 8661 (15.3.4); 8662 (5.4.1); 8663 (26.13.15); 8664 (28.6.26); 8665 (26.7.3); 8666 (10.1.5); 8667 (27.3.6); 8668

(23.4.2); 8669 (16.16.2); 8670 (16.10.5); 8671 (16.14.1); 8672 (16.7.3); 8673 (16.10.1); 8674 (16.4.3); 8675 (16.4.1); 8676 (16.8.2); 8677 (16.8.2); 8678 (28.2.2); 8679 (16.12.2); 8680 (16.7.3); 8681 (16.7.2); 8682 (16.8.11); 8683 (16.10.3); 8684 (16.10.3); 8685 (16.10.4); 8686 (16.8.9); 8687 (16.8.3); 8688 (16.8.2); 8689 (17.2.2); 8690 (23.4.1); 8691 (6.1.31); 8692 (16.8.4); 8693 (16.8.2); 8694 (16.7.3); 8695 (16.8.7); 8696 (26.13.8); 8697 (28.2.2); 8698 (14.3.3a); 8699 (14.3.1); 8700 (15.2.6); 8701 (15.4.13); 8702 (15.3.8); 8703 (7.1.4); 8704 (13.13.2); 8705 (16.7.4); 8706 (16.7.4); 8707 (16.8.10); 8708 (16.8.10); 8709 (15.4.13); 8710 (17.2.1); 8711 (16.8.10); 8712 (16.8.10); 8713 (16.10.3); 8714 (23.2.3a); 8715 (15.4.8); 8716 (15.2.11); 8717 (15.3.8); 8718 (15.3.4); 8719 (15.2.3); 8720 (16.12.2); 8721 (15.4.9); 8722 (15.4.9); 8723 (15.4.29); 8724 (16.10.1); 8725 (15.6.4); 8726 (17.2.1); 8727 (23.16.2); 8728 (6.1.10); 8729 (26.4.2a); 8730 (25.2.3); 8731 (16.8.2); 8732 (16.4.3); 8733 (16.8.10); 8734 (16.5.1); 8735 (16.8.10); 8736 (16.7.4); 8737 (15.3.14); 8738 (15.2.10); 8739 (28.2.3); 8740 (15.3.14); 8741 (13.13.2); 8742 (13.8.1); 8743 (6.1.2); 8744 (15.4.28); 8745 (22.1.3); 8746 (22.1.2); 8747 (22.1.4); 8748 (16.5.1); 8749 (16.8.10); 8750 (15.4.17); 8751 (15.4.28); 8752 (15.3.14); 8753 (15.4.13); 8754 (15.6.2a); 8755 (15.6.2a); 8756 (16.5.1); 8757 (15.4.3); 8758 (15.4.27); 8759 (15.4.15); 8760 (15.4.13); 8761 (16.8.3); 8762 (15.4.15); 8763 (15.4.27); 8765 (15.4.13); 8766 (13.8.1); 8767 (23.2.3a); 8768 (16.4.3); 8769 (6.1.13); 8770 (6.1.29); 8771 (6.1.13); 8772 (13.13.3); 8773 (13.13.3); 8774 (6.1.29); 8775 (16.8.10); 8776 (16.8.10); 8777 (16.4.3); 8778 (13.15.1); 8779 (28.2.3); 8780 (13.1.1); 8781 (10.3.1); 8782 (15.4.8); 8783 (15.3.14); 8784 (15.2.6); 8785 (6.1.29); 8786 (10.7.1); 8787 (8.1.9); 8788 (15.4.9); 8789 (15.6.2a); 8790 (8.1.7); 8791 (9.4.4); 8792 (16.4.3); 8793 (26.4.2a); 8794 (6.1.29); 8795 (16.8.4); 8796 (15.3.14); 8797 (15.2.3); 8798 (15.2.6); 8799 (16.8.2); 8800 (7.1.10); 8801 (14.2.1); 8802 (13.9.4); 8803 (23.16.2); 8804 (8.1.9); 8805 (17.2.4); 8806 (17.2.7); 8807 (9.4.4); 8808 (23.4.5); 8809 (12.2.2); 8810 (14.2.2); 8811 (7.1.5); 8812 (28.2.1); 8813 (28.6.31); 8814 (8.1.13); 8815 (13.11.2); 8816 (17.2.1); 8817 (7.1.1); 8818 (14.3.1); 8819 (10.7.1); 8820 (10.3.2); 8821 (28.6.26); 8822 (16.7.2); 8823 (15.2.6); 8824 (15.2.2); 8825 (15.4.14); 8826 (15.2.11); 8827 (15.2.11); 8828 (15.2.11); 8829 (15.2.10); 8830 (15.2.2); 8831 (15.2.6); 8832 (15.2.11); 8833 (15.2.6); 8834 (14.4.3a); 8835 (28.6.20); 8836 (15.4.29); 8837 (14.4.3a); 8838 (16.8.2); 8839 (16.8.4); 8840 (16.8.8); 8841 (16.10.2); 8842 (8.1.9); 8843 (10.8.1); 8844 (10.9.1a); 8845 (12.2.1); 8846 (12.2.1); 8847 (8.1.14); 8848 (17.2.3); 8849 (16.12.2); 8850 (16.12.2); 8851 (15.4.11); 8852 (17.2.6); 8853 (17.2.1); 8854 (23.16.4); 8855 (23.16.3a); 8856 (14.1.3a); 8857 (14.4.1a); 8858 (14.4.4a); 8859 (14.2.1); 8860 (24.1.6); 8861 (28.2.5); 8862 (28.2.4); 8863 (26.2.2a); 8864 (26.13.14); 8865 (26.7.3); 8866 (10.5.2); 8867 (13.3.3); 8868 (10.3.2); 8869 (6.1.21); 8870 (15.2.3); 8871 (15.2.1); 8872 (28.1.1); 8873 (20.3.3); 8874 (28.3.1); 8875 (28.3.1); 8876 (15.3.5); 8877 (15.3.10); 8878 (8.1.3); 8879 (16.3.2); 8880 (16.4.4); 8881 (6.1.34); 8882 (6.1.11); 8883 (10.4.13); 8884 (23.3.1); 8885 (28.6.26); 8886 (28.6.7b); 8887 (6.1.16); 8888 (9.4.3); 8889 (10.4.18); 8890 (10.4.14); 8891 (13.9.1); 8892 (15.2.4); 8893 (15.2.4); 8894 (15.2.3); 8895 (15.2.2); 8896 (15.2.4); 8897 (15.2.4); 8898 (15.2.4); 8899 (26.11.6); 8900 (23.3.3); 8901 (23.13.4); 8902 (26.13.13); 8903 (24.1.11); 8904 (17.3.2); 8905 (6.1.34); 8906 (6.1.33); 8907 (6.1.36); 8908 (23.6.1); 8910 (14.4.4a); 8911 (13.10.1); 8912 (23.22.1); 8913 (28.6.9); 8914 (27.1.8); 8916 (6.1.30); 8917 (13.16.10); 8918 (16.9.2); 8919 (16.3.2); 8920 (6.1.23); 8921 (26.3.2); 8922 (26.11.7b); 8923 (11.2.1); 8924 (14.4.2); 8925 (13.4.4); 8926 (5.1.2); 8927 (7.1.2); 8928 (5.6.1); 8929 (16.2.2); 8930 (9.4.3); 8931 (26.8.1a); 8932 (16.4.3); 8933 (9.2.3); 8934 (26.13.4); 8935 (15.6.3); 8936 (10.4.8); 8937 (13.16.5); 8938 (10.10.4); 8939 (10.10.3a); 8940 (13.4.1); 8941 (13.4.1); 8942 (13.4.3a); 8943 (15.6.3); 8944 (24.1.4); 8945 (23.20.3); 8946 (28.6.8); 8947 (24.1.17); 8948 (28.6.21); 8949 (26.10.2); 8950 (6.1.19b); 8951 (10.5.4a); 8952 (6.1.36); 8953 (26.3.2); 8954 (27.3.2); 8955 (26.11.4); 8956 (28.6.1); 8957 (8.1.8); 8958 (15.1.2); 8959 (10.10.3a); 8960 (5.7.1); 8961 (10.4.15); 8962 (10.4.15); 8963 (10.4.13); 8964 (10.4.2); 8965 (27.1.8); 8966 (23.3.4); 8967 (9.4.2); 8968 (26.13.4); 8969 (16.9.1); 8970 (16.16.1); 8971 (16.3.4); 8972 (16.3.2); 8973 (16.11.1); 8974 (16.10.6); 8975 (16.7.5); 8976 (16.8.13); 8977 (16.8.2); 8978 (16.8.5); 8979 (16.8.9); 8980 (14.1.3d); 8981 (6.1.34); 8982 (6.1.30); 8983 (23.20.4); 8984 (6.1.35); 8985 (23.20.3); 8986 (6.1.17); 8987 (23.15.1); 8988 (9.2.1); 8989 (23.8.4); 8990 (18.1.3); 8991 (11.3.1); 8992 (16.14.1); 8993 (16.2.2); 8994 (28.6.19); 8995 (16.3.2); 8996 (20.2.1); 8997 (20.2.5); 8998 (13.16.2); 8999 (13.12.1); 9000 (8.1.8); 9001 (26.13.4); 9002 (13.10.2); 9003 (23.13.10); 9004 (26.13.11); 9005 (23.7.1); 9006 (26.1.1); 9007 (23.12.1); 9008 (8.1.4); 9009 (23.13.1); 9010 (24.1.14); 9011 (9.3.1); 9012 (6.1.24); 9013 (23.13.7); 9014 (26.13.13); 9015 (23.11.5); 9016 (23.11.2); 9017 (6.1.36); 9018 (6.1.14); 9019 (15.4.18); 9020 (15.4.4); 9021 (15.6.4); 9022 (15.2.11); 9023 (15.3.12); 9024 (15.7.1); 9025 (16.8.11); 9026 (15.3.18); 9027 (6.1.8); 9028 (24.1.5); 9029 (15.3.15); 9030 (20.2.3); 9031 (20.2.3); 9032 (13.9.5); 9033 (15.4.4); 9034 (18.1.4); 9035 (6.1.22); 9036 (13.9.6); 9037 (28.6.29a); 9038 (10.3.3); 9039 (14.1.2); 9040 (15.4.24); 9041 (15.2.5); 9042 (16.7.6); 9043 (16.3.2); 9044 (16.16.2); 9045 (17.2.6); 9046 (17.5.1); 9047 (23.16.3a); 9048 (26.11.3); 9049 (26.13.3); 9050 (26.13.2); 9051 (26.13.2); 9052 (10.4.2); 9053 (10.4.8); 9054 (10.4.4); 9055 (10.4.10); 9056 (10.4.13); 9057 (20.2.5); 9058 (10.4.12); 9059 (11.1.1); 9060 (16.1.1); 9061 (15.3.6); 9062 (15.3.5); 9063 (15.4.5); 9064 (15.4.4); 9065 (15.4.20); 9066 (15.4.9); 9067 (15.4.14); 9068 (15.6.4); 9069 (15.3.7); 9070 (15.2.1); 9071 (15.4.14); 9072 (15.4.18); 9073 (15.4.9); 9074 (15.6.2b); 9075 (13.3.3); 9076 (10.4.4); 9077 (6.1.10); 9078 (10.4.12); 9079 (26.7.3); 9080 (26.13.14); 9081 (9.5.1); 9082 (23.8.4); 9083 (24.1.11); 9084 (26.12.3); 9085 (26.12.3); 9086 (13.16.4); 9087 (26.2.1); 9088 (13.9.1); 9089 (13.13.4); 9090 (13.13.1); 9091 (23.13.3); 9092 (10.4.18); 9093 (26.13.5); 9094 (28.4.3a); 9095 (24.1.8); 9096 (24.1.6); 9097 (24.1.18); 9098 (6.1.7); 9099 (28.5.1); 9100 (10.5.2); 9101 (26.10.4); 9102 (26.13.1); 9103 (28.6.24); 9104 (28.6.13); 9105 (26.12.1); 9106 (26.7.2); 9107 (10.1.4a); 9108 (10.1.4a); 9109 (10.3.2); 9110 (6.1.22); 9111 (17.2.2);

9112 (6.1.16); 9113 (6.1.16); 9114 (6.1.16); 9115 (15.4.24); 9116 (15.2.5); 9117 (15.4.18); 9118 (15.4.4); 9119 (26.6.2); 9120 (26.11.6); 9121 (21.3.2b); 9122 (28.6.11); 9123 (15.2.8); 9124 (14.2.3); 9125 (26.7.1); 9126 (26.11.2); 9127 (14.2.4); 9128 (23.18.1); 9129 (17.5.1); 9130 (28.6.7b); 9131 (28.6.7a); 9132 (28.6.7a); 9133 (23.16.4); 9134 (27.3.1); 9135 (13.2.1); 9136 (6.1.9); 9137 (15.5.2); 9138 (5.1.1); 9139 (16.7.8); 9140 (16.8.4); 9141 (16.8.11); 9142 (23.13.5); 9143 (23.16.1); 9144 (6.1.16); 9145 (26.13.9); 9146 (13.13.1); 9147 (23.13.11); 9148 (13.9.5); 9149 (15.3.7); 9150 (16.14.1); 9151 (6.1.37a); 9152 (17.2.5); 9153 (6.1.16); 9154 (28.6.21); 9155 (23.1.1); 9156 (24.1.14); 9157 (16.7.7); 9158 (16.15.1); 9159 (16.4.4); 9160 (16.8.4); 9161 (16.7.4); 9162 (16.12.2); 9163 (16.7.3); 9164 (16.12.1); 9166 (15.2.8); 9167 (15.2.2); 9168 (15.4.4); 9169 (15.4.16); 9170 (15.3.12); 9171 (15.3.21); 9172 (15.7.1); 9173 (15.7.1); 9174 (15.7.1); 9175 (23.20.3); 9176 (16.10.5); 9177 (10.4.13); 9178 (23.1.1).

Poore 408 (14.3.3a); 428 (15.4.4).

Price 138 (15.3.4); 207 (7.1.9); 238 (2.3.1); 239 (15.2.3).

RSNB (less than 2980) under Chew, Corner & Stainton; RSNB (more than 4000) under Chew & Corner

Saikeh 82754 (2.3.2).

Shea & Aban 76864 (23.6.1); 76905 (23.13.6); 76906 (26.11.6); 76909 (27.1.7); 76911 (26.13.11); 77117 (5.5.1); 77133 (23.5.1); 77159 (26.11.6); 77177 (24.1.9); 77185 (25.1.1); 77189 (20.2.1); 77191 (26.3.3); 77222 (28.6.21); 77239 (10.10.4); 77241 (26.13.4); 77265 (23.9.2); 77266 (9.2.1); 77269 (26.13.16a); 77280 (23.21.1); 77282 (6.1.24); 77313 (18.1.3); 77321 (8.1.4).

Shim 75401 (23.8.3); 75402 (15.4.19); 75408 (17.2.5); 75421 (15.4.4); 75444 (15.3.6); 75447 (15.2.3); 75458 (15.7.1); 81801 (10.9.1a).

Sinclair et al. 8960 (1.1.1a); 8970 (16.8.1); 8972 (6.1.21); 8978 (16.8.3); 8981 (16.8.6); 8982 (10.4.17a); 8983 (8.1.16); 8984 (8.1.21); 8999 (11.2.1); 9005 (11.4.1); 9008 (15.2.3); 9014 (16.15.1); 9018 (16.6.1); 9026 (27.3.1); 9028 (15.2.1); 9032 (10.4.18); 9039 (14.3.4); 9040 (7.1.9); 9046 (2.3.5); 9047 (26.4.2a); 9048 (22.1.1); 9054 (14.4.1a); 9055 (23.16.2); 9055A (23.4.1); 9055a (23.4.1); 9067 (14.3.3a); 9070 (16.12.2); 9078 (23.4.5); 9095 (14.2.2); 9100 (25.2.3); 9106 (13.13.2); 9107 (15.4.8); 9117 (6.1.10); 9119 (23.16.2); 9121 (22.1.2); 9139 (2.3.2); 9151 (7.1.4); 9152 (15.4.13); 9154 (23.2.3a); 9170 (28.2.1); 9171 (13.13.2); 9183 (13.8.1); 9183A (13.8.1); 9190 (10.2.1); 9192 (8.1.9); 9193 (8.1.7); 9196 (2.3.4); 9202 (14.3.4); 9205 (13.3.2); 9207 (16.7.6); 9222 (16.1.1); 9229 (10.5.2); 9231 (8.1.4); 9232 (10.3.3); 9234 (8.1.17).

Smith 530 (15.3.14); 532 (15.2.10); 533 (15.4.17).

Tamura & Hotta 309 (6.1.19a); 356 (18.1.5); 358 (23.13.9); 371 (23.13.11); 507 (23.13.2).

Topping 1504 (10.4.11a); 1505 (24.1.1); 1511 (23.8.4); 1513 (26.3.3); 1514 (11.1.1); 1515 (10.8.1); 1516 (14.1.3d); 1518 (8.1.4); 1520 (11.1.1); 1531 (26.13.11); 1536 (10.4.17c); 1537 (28.6.15); 1540 (13.16.2, 13.16.5); 1545 (6.1.34); 1550 (28.6.7a); 1552 (27.3.4); 1556 (23.13 1); 1558 (24.1.4); 1560 (26.3.1); 1567 (10.4.17c); 1568 (24.1.5); 1569 (10.1.2); 1572 (23.3.3); 1573 (10.4.8); 1574 (23.13.9); 1581 (9.3.1); 1587 (28.4.1); 1588 (23.22.1); 1598 (6.1.9); 1600 (16.3.2); 1602 (24.1.2); 1605 (24.1.3); 1607 (6.1.34); 1608 (5.1.1); 1612 (27.3.7); 1613 (24.1.11); 1619 (16.8.12); 1625 (12.1.1); 1630 (20.2.3); 1631 (28.6.7b); 1632 (10.5.1); 1633 (14.2.4); 1636 (21.3.3a); 1641 (16.7.2); 1643 (10.10.3a); 1656 (23.4.5); 1659 (15.2.2); 1663 (15.4.6); 1665 (15.4.8); 1666 (15.2.2); 1668 (15.4.8); 1671 (10.3.3); 1675 (23.2.3a); 1677 (6.1.10); 1679 (15.5.3); 1696 (13.15.1); 1698 (28.2.3); 1702 (13.13.3); 1703 (23.2.3a); 1704 (13.13.2); 1711 (15.4.8); 1713 (15.2.2, 15.2.10); 1714 (15.3.8); 1715 (15.5.3); 1716 (28.2.1, 28.2.4); 1719 (26.4.2a); 1720 (25.2.3); 1721 (16.8.5); 1724 (25.2.3); 1726 (15.2.2); 1734 (23.4.5); 1737 (6.1.10); 1745 (9.4.4); 1749 (23.4.1); 1750 (12.1.1); 1752 (16.16.2); 1753 (14.4.1a); 1755 (14.4.3a); 1756 (28.6.20); 1758 (8.1.7); 1759 (8.1.11); 1762 (26.5.2); 1766 (26.13.9); 1770 (15.3.5); 1771 (16.7.2); 1773 (6.1.37a); 1774 (23.3.1); 1777 (6.1.22); 1778 (23.13.5); 1780 (9.2.2); 1782 (10.1.2); 1784 (28.6.9); 1788 (26.13.3, 26.13.10a); 1793 (6.1.34); 1794 (13.13.1); 1796 (6.1.25); 1798 (6.1.3); 1800 (25.1.1); 1802 (23.13.1); 1804 (8.1.4); 1805 (23.1.1); 1806 (16.3.2); 1807 (10.4.17c); 1813 (16.3.1); 1814 (6.1.16); 1815 (6.1.3); 1816 (15.2.3); 1820 (10.4.12); 1823 (23.18.1); 1824 (8.1.16); 1833 (27.3.7); 1835 (28.6.7b); 1837 (26.13.5); 1841 (10.4.17c); 1842 (9.4.4); 1843 (28.6.8); 1847 (16.14.1); 1850 (8.1.10); 1850½ (26.13.7); 1851 (10.4.1); 1856 (20.3.2); 1862 (6.1.5); 1866 (10.4.8); 1868 (5.6.1); 1870 (14.3.2); 1871 (10.4.12); 1877 (16.14.1); 1879 (16.6.1); 1881 (16.8.5); 1889 (21.3.3a); 1890 (10.4.9); 1894 (25.1.1); 1895 (23.8.4); 1896 (6.1.35); 1899 (26.1.1); 1902 (26.13.12); 1907 (10.5.4a); 1908 (27.3.7); 1910 (13.16.2); 1959 (8.1.11).

de Vogel 8014 (23.16.4).

Weber 54729 (2.3.2).

APPENDIX

While this treatment was in press Professor K.U. Kramer examined it and made several important suggestions, some of which were possible to incorporate. Some other corrections are recorded in this appendix, along with a number of changes resulting from Parris's work with 'Families and Genera of Vascular Plants' which will be published by The Royal Botanic Gardens, Kew in 1992.

The addition of two new families brings the number of families included in this treatment to 30. The reduction to synonymy of seven genera in Polypodiaceae brings the number of genera to 138. The total number of species included in the treatment is reduced to 608 and the number of taxa to 620. The number of new combinations for species published here by Parris is increased from 10 to 21. The enumeration now includes 28 rather than 29 taxa that will be described later by Parris.

10.6.1. Orthiopteris kingii becomes **Saccoloma kingii** (Bedd.) Parris, comb. nov., *Dicksonia kingii* Bedd., Handb. Suppl., 6 (1892).

10.9.1. Sphenomeris chinensis becomes **Odontosoria chinensis** (L.) J. Sm., Bot. Voy. Herald, 430 (1857).

10.9.2. Sphenomeris retusa becomes **Odontosoria retusa** (Cav.) J. Sm., Bot. Voy. Herald, 430 (1857).

10.9.3. Sphenomeris veitchii becomes **Odontosoria veitchii** (Baker) Parris, comb. nov., *Davallia veitchii* Baker, J. Bot. 17: 39 (1879).

11.2.1. Culcita javanica becomes **Calochlaena javanica** (Blume) R.A. White & M.D. Turner, Amer. Fern J. 78: 93 (1988).

12.1. Cheiropleuria is to be removed from Dipteridaceae to its own family Cheiropleuriaceae.

13.6.1. Diacalpe aspidioides becomes **Peranema aspidioides** (Blume) Mett., Fil. Lechl. 2: 33 (1859).

20.2 Nephrolepis is to be removed from Oleandraceae to its own family Nephrolepidaceae.

23.4.1. Crypsinus albidopaleatus becomes **Selliguea albidopaleata** (Copel.) Parris, comb. nov., *Polypodium albidopaleatum* Copel., Philipp. J. Sci. 12C: 63 (1917).

23.4.2. Crypsinus enervis becomes **Selliguea enervis** (Cav.) Ching, Bull. Fan Mem. Inst. Biol. Bot. 10: 239 (1941).

23.4.3. Crypsinus platyphyllus becomes **Selliguea platyphylla** (Sw.) Ching, Bull. Fan Mem. Inst. Biol. Bot. 10: 238 (1941).

23.4.4. Crypsinus stenophyllus becomes **Selliguea stenophylla** (Blume) Parris, comb. nov., *Polypodium stenophyllum* Blume, Enum. Pl. Javae, 124 (1828).

23.4.5. Crypsinus stenopteris becomes **Selliguea stenopteris** (Baker) Parris, comb. nov., *Polypodium stenopteris* Baker, J. Bot. 17: 43 (1879).

23.4.6. Crypsinus taeniophyllus becomes **Selliguea taeniophylla** (Copel.) Parris, comb. nov., *Polypodium taeniophyllum* Copel., Phillip. J. Sci. 7C: 65 (1912).

23.5.1. Drymoglossum piloselloides becomes **Pyrrosia piloselloides** (L.) M.G. Price, Kalikasan 3: 176 (1974).

23.7.1. Drynariopsis heraclea becomes **Aglaomorpha heraclea** (Kunze) Copel., Univ. Calif. Publ. Bot. 16: 117 (1929).

23.12.1 Merinthosorus drynariodes becomes **Aglaomorpha drynariodes** (Hook.) Roos, Blumea 31: 153 (1985).

23.13. Add to literature: Bosman, M.T.M. 1991. A monograph of the fern genus *Microsorum* (Polypodiaceae). Leiden Bot. Ser. 14.

23.14.1. Paragramma longifolia becomes **Lepisorus longifolius** (Blume) Holttum, Rev. Fl. Malaya 2, ed. 2: 151 (1968).

23.15.1. Photinopteris speciosa becomes **Aglaomorpha speciosa** (Blume) Roos, Blumea 31: 153 (1985).

23.16.1. Phymatopteris albidosquamata becomes **Selliguea albidosquamata** (Blume) Parris, comb. nov., *Polypodium albidosquamatum* Blume, Enum. Pl. Javae, 132 (1828).

23.16.2. Phymatopteris pakkaensis becomes **Selliguea pakkaensis** (C. Chr.) Parris, comb. nov., *Polypodium pakkaense* C. Chr. in C. Chr. & Holttum, Gardens' Bull. 7: 310 (1934).

23.16.3. Phymatopteris taeniata becomes **Selliguea taeniata** (Sw.) Parris, comb. nov., *Polypodium taeniatum* Sw., Schrad. J. Bot. 1800 (2): 26 (1801).

23.16.3.b. Phymatopteris taeniata var. **palmata** (Blume) Parris, comb. nov. becomes **Selliguea taeniata** var. **palmata** (Blume) Parris, comb. nov., *Polypodium palmatum* Blume, Fl. Javae 2: 150 (1829).

23.16.4 Phymatopteris triloba becomes **Selliguea triloba** (Houtt.) M.G. Price, Contr. Univ. Mich. Herb. 17: 276 (1990). *Selliguea triloba* should of course be deleted from the synonymy.

23.19.1. Pycnoloma metacoelum becomes **Selliguea metacoela** (Alderw.) Parris, comb. nov., *Drymoglossum metacoelum* Alderw., Bull. Jard. Bot. Buit. 2, 28: 21 (1918).

23.19.2. Pycnoloma murudense becomes **Selliguea murudensis** (C.Chr.) Parris, comb. nov., *Pycnoloma murudense* C. Chr., Dansk Bot. Arkiv. 6: 78, t. 8, 10 (1929).

24.1.8. Pteris kinabaluensis is synonymized with **P. longipes** D. Don.

24.1.12. Pteris purpureorachis is synonymized with **P. amoena** Blume, Enum. Pl. Javae, 210 (1828).

24.1.14. The specimens *Clemens 32562, 34142,* and *Parris & Croxall 9156* cited under **Pteris tripartita** are **P. pediformis** Kato & Kramer, Acta Phytotax. Geobot. 41: 166, f. 4, (1990).

24.1.17. The specimen (*Parris & Croxall 8947*) included as **Pteris sp. 1** is **P. longipes** D. Don.

24.1.18. Pteris sp. 2 becomes **Pteris sp. 1. 24.1.18** becomes **24.1.17**.

INDEX TO SCIENTIFIC NAMES

Accepted names are in roman type. Synonyms are in *italics*.

Athyrium continued
  amoenum C. Chr. in C. Chr. & Holttum,
    129
  anisopterum Christ, 129
  atropurpureum Copel., 56
  atrosquamosum Copel., 131
  clemensiae Copel., 129
  macrocarpum sensu C. Chr. & Holttum non
    (Blume) Bedd., 129
  megistophyllum Copel., 134
  pulcherrimum Copel., 129
  sp. 1, 129

Belvisia, 10, 98
  mucronata (Fée) Copel., 98
  spicata (L. f.) Mirb. ex Copel., 98
  squamata (Hieron. ex C. Chr.) Copel., 98
    var borneensis (C. Chr.) Parris, comb.
      nov., 98
Blechnaceae, 6, 29
Blechnum, 6, 9, 29
  borneense C. Chr., 29
  egregium Copel., 29
  finlaysonianum Wall. ex Hook. & Grev., 29
  fluviatile (R. Br.) Lowe ex Salomon, 6, 9, 29
  fraseri (Cunn.) Luerss., 29
    var. philippinense (Christ) C. Chr. in C.
      Chr. & Holttum, 30
  orientale L., 6, 30
  pallescens T. C. Chambers, ined., 30
  patersonii (R. Br.) Mett., 30
  procerum sensu C. Chr. & Holttum non (G.
    Forster) Sw., 30
  sp. 1, 30
  sp. 2, 31
  vestitum (Blume) Kuhn, 30
  vulcanicum (Blume) Kuhn, 30
Bolbitis, 89
  repanda (Blume) Schott, 89
  sinuata (C. Presl) Hennipman, 89
Botrychium, 96
  daucifolium Wall. ex Hook. & Grev., 96

Callistopteris, 11, 77
  apiifolia (C. Presl) Copel., 77
Calochlaena javanica (Blume) R.A. White &
  M.D. Turner, 151
Calymmodon, 62
  clavifer (Hook.) T. Moore, 62
  cucullatus (Nees & Blume) C. Presl, 62
  gracilis (Fée) Copel., 62
  hygroscopicus Copel., 63
  muscoides (Copel.) Copel., 2, 63
  sp. 1, 63
  sp. 2, 63
  sp. 3, 63
  sp. 4, 2, 63
  sp. 5, 64
  sp. 6, 64
  sp. 7, 2, 64
Campium
  quoyanum sensu C. Chr. & Holttum non
    (Gaud.) Copel., 89
  subsimplex (Fée) Copel., 89
Cephalomanes, 11, 77
  javanicum (Blume) Bosch, 77
  laciniatum (Roxb.) De Vol in H. L. Li et al.,
    78

Cheilanthes, 21
  tenuifolia (N. L. Burm.) Sw., 21
Cheiropleuria, 48, 151
  bicuspis (Blume) C. Presl, 8, 48
Cheiropleuriaceae, 8, 151
Chingia, 115
  atrospinosa (C. Chr.) Holttum, 115
  clavipilosa Holttum, 115
    var. clavipilosa, 115
Christella, 116
  arida (D. Don) Holttum in Nayar & Kaur,
    116
  hispidula (Decne.) Holttum, 116
  parasitica (L.) Levier, 116
  subpubescens (Blume) Holttum, 116
Christensenia, 93
  aesculifolia (Blume) Maxon, 93
Cibotium, 46
  arachnoideum (C. Chr.) Holttum, 7, 47
  cumingii Kunze
    var. arachnoideum C. Chr. in C. Chr. &
      Holttum, 47
Colysis, 99
  acuminata Holttum, 99
  loxogrammoides (Copel.) M. G. Price, 99
  macrophylla (Blume) C. Presl, 99
  pedunculata (Hook. & Grev.) Ching, 99
Coniogramme, 21
  fraxinea (D. Don) Diels, 21
  macrophylla (Blume) Hieron., 21
    var. copelandii (Christ) Hieron., 21
Cornopteris, 130
  opaca (D. Don) Tagawa, 130
Coryphopteris, 8, 116
  badia (Alderw.) Holttum, 116
  gymnopoda (Baker) Holttum, 117
    var. gymnopoda, 117
  multisora (C. Chr.) Holttum, 117
  obtusata (Alderw.) Holttum, 117
Crepidomanes, 11, 78
  bilabiatum (Nees & Blume) Copel., 78
  bipunctatum (Poir.) Copel., 78
  brevipes (C. Presl) Copel., 78
  christii (Copel.) Copel., 78
Crypsinus, 99
  albidopaleatus (Copel.) Copel., 100, 151
  enervis (Cav.) Copel., 100, 151
  platyphyllus (Sw.) Copel., 100, 151
  stenophyllus (Blume) Holttum, 100, 151
  stenopteris (Baker) Parris, 100, 152
  taeniophyllus (Copel.) Copel., 101, 152
Ctenitis, 50
  aciculata (Baker) Ching, 50
  atrorubens Holttum, 50
  kinabaluensis Holttum
    var. kinabaluensis, 50
  minutiloba Holttum, 50
Ctenopteris, 4, 64
  barathrophylla (Baker) Parris, 64
  aff. barathrophylla (Baker) Parris, 64
  blechnoides (Grev.) W. H. Wagner &
    Grether, 65
  brevivenosa (Alderw.) Holttum, 65
  celebica (Blume) Copel., 65
  curtisii (Baker) Copel., 65
  denticulata (Blume) C. Chr. & Tardieu-Blot,
    2, 65
  fuscata (Blume) Kunze, 66